Getting Started with UVM
A Beginner's Guide

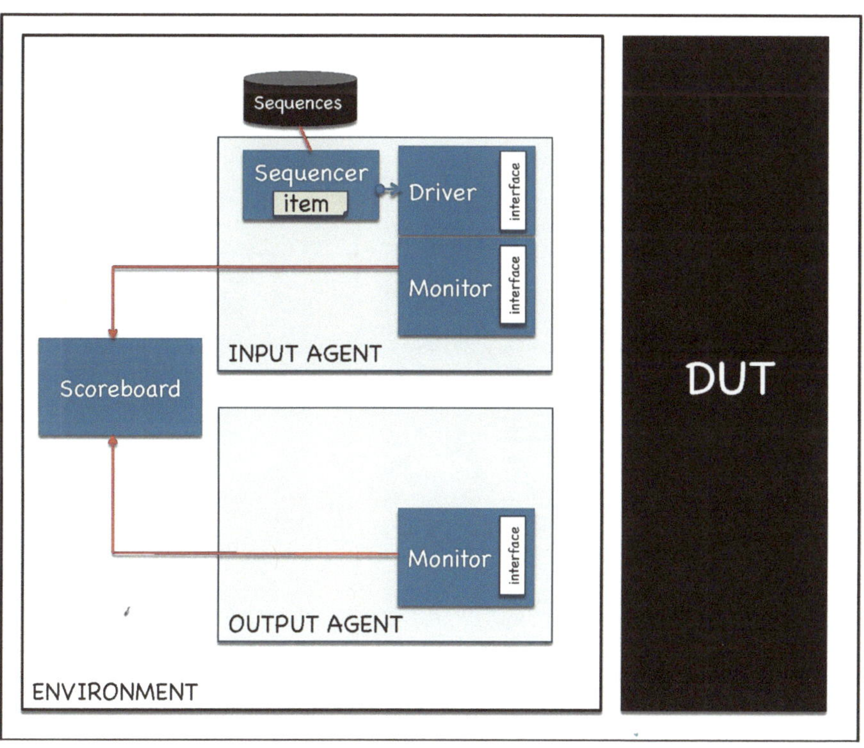

Vanessa R. Cooper

verilab

Project Manager: **JL Gray**
Designer: **Vanessa R. Cooper**
Editor: **Georgellen Burnett**

Verilab Publishing is an imprint of Verilab, Inc.
8310 North Capital of Texas Highway
Suite 215
Austin, TX 78731, USA

ISBN-13: 978-0615819976 (Verilab Publishing)
ISBN-10: 0615819974

About Verilab, Inc.

Verilab specializes in verification environment development, methodology definition, and project planning for chip design projects. Since our founding in 2000, we have worked on over 200 projects across more than 50 companies. With offices in the US, UK, Germany, and Canada, our goal is to facilitate consistent, successful results for our clients.

SystemVerilog + UVM/OVM/VMM

Verilab is involved with the IEEE-1800 (SystemVerilog) committee and, as an Accellera member, an active participant in the VIP-TSC (UVM). Our consultants have access to up-to-date technical information on the latest methodologies and trends.

Specman/e + eRM/UVMe

Looking for a seasoned Specmaniac to join your project team? Many of our consultants have years of experience bulding verification environments in e.

Verification Leadership

Verilab has helped companies big and small in both direct leadership and consulting roles to help navigate the minefield between design concept and tapeout.

Clock Domain Crossing Design and Verification

Verilab has helped companies around the world do a better job understanding CDC design and verification best practices.

SystemRDL, UVM-ML

Verilab engineers are keeping an eye on emerging Accellera standards in the areas of register modeling and multi-language verification methodologies.

How We Work With Clients

Onsite

Our engineers can work directly with your project team onsite at your office.

Remote/Onsite

When our engineers are located remotely from your facility, we recommend they come onsite for 1-2 weeks for a project kickoff. Typically, they would come onsite for one week per month for the duration of the project.

Consulting

Sometimes you have the right team in place but need some additional knowledge and guidance to get the job done. A Verilab consultant can come onsite for 1-2 weeks for an intensive ramp up on topics such as verification planning, SystemVerilog/UVM testbench development, or CDC design and verification. We would then follow up with you over the next several months either via short phone calls or occasional onsite visits to check on progress.

Leadership Coaching

Have you recently moved into a verification leadership position? Work with a senior member of Verilab's engineering team for direct 1:1 coaching, either via phone or in person, to guide you through your first project.

For more information on how we can best serve you, contact sales@verilab.com.

About the Author

Vanessa Cooper has thirteen years of experience as both a digital design verification and support applications engineer. She has worked on products ranging from automotive applications to consumer electronics.

Since joining Verilab, Inc. in 2011 as a verification engineer, she has implemented verification environments in UVM and has worked with OVM.

Vanessa holds a B.S. in Computer Engineering from Mississippi State University and an M.B.A. from LeTourneau University.

Foreword

There are a multitude of books out there to help you hone your Verification skills, as almost every Digital IC professional will tell you. The available books run the gamut of quality and ease of use. The major strength of this book is a place where many of those books fall down – they are not what they claim to be. Vanessa's book is a great example of exactly what it claims to be – a guide for the "beginner". "Beginner" is used loosely in this context because the book does not shy away from advanced concepts (the Factory, register package prediction, etc.).

Her common-sense approach and liberal use of examples are great for quickly building a framework for understanding UVM. She walks the reader through the methodology in nice, linear fashion, and does not waste time explaining every single constrained-random Verification concept. This book is intended to be a practical guide for getting a D/V engineer started with UVM, and it succeeds. The book explains important concepts such as macros, messaging, and the Factory well without dwelling overly long on underlying details. The focus is kept on day-to-day useful information. This book is a great resource, regardless of whether you're used to a Verilog task-based approach or another SystemVerilog-based methodology (although I recommend you study SystemVerilog or Object-Oriented concepts a bit before using any SV methodology).

The book is also quite thorough and complete in presenting its subject. As a Verification engineer, I didn't find myself grasping for that next essential piece of information. This book should not require the use of web searches to comprehend its key concepts, despite the concept's simplicity or complexity (though perusal of http://www.accellera.org/community/uvm is recommended, a point with which I agree). The reader should be developing fully functional UVM testbenches after in-depth study of this relatively short tome. Hopefully, these few words will spur you towards giving this book a try. It really is a solid resource for Digital IC folks who are trying to "make the jump" to UVM. It is a book I'm glad I have on my shelf.

Ray Harlan, Cirrus Logic

Acknowledgments

Writing this book has been a great joy, and I appreciate the support and encouragement of my fellow Verilab colleagues. A special thank you to those who took the time to review the material:

JL Gray
Jason Sprott
Gordon McGregor
Jonathan Bromley
Mark Litterick
Jeff Montesano
Alex Melikian
Taruna Reddy

Also, I owe a tremendous debt of gratitude to those outside of Verilab who gave of their time to review the book:

Mark Glasser, Cypress Semiconductor
Andrew Stortz, Intel, Inc.
Irina Furman, Intel, Inc.
Ray Harlan, Cirrus Logic
Asif Jafri, ARM

Finally, this book would not have been possible without the support of my wonderful husband, Doug, and my loving son, Lucas. Thanks guys!

Chapter 1
Introduction

You are just starting a new project or perhaps a derivative of the chip you are on is on the verge of taping out. Your manager has just informed the team that for this new project you will adopt the UVM methodology. You think, "Great ... yet another methodology!" Now, you have to learn it and get started in a fairly short period of time. Perhaps you have taken the UVM course from your EDA vendor of choice, or perhaps you haven't. Either way, you are sitting in front of your Linux terminal wondering, "How do I begin?" This book is for you.

The goal of this book is to help you get up and going quickly with UVM. The User and Reference guides that are available with the installation are excellent resources. I also recommend viewing the actual source code. Reading the source code will give you a better understanding of how the library is implemented. You can find the source code in the installation of your simulation tool or from the UVM World website which links to Accellera: http://www.accellera.org/community/uvm.

This book is meant to serve as an additional resource. My hope is that it provides you with succinct information and meaningful examples that relate to your own environment. I also hope to provide you with some "gotchas" that I have come across while learning UVM so that you can avoid similar mistakes. Since UVM is a library for use with SystemVerilog verification environments, this book assumes a preexisting knowledge of SystemVerilog.

Before we dive into the "how to begin" details, let's do some housekeeping with the "before you begin" necessities. Let's briefly explore some of the advantages of UVM and planning.

Advantages of the Universal Verification Methodology

The main advantages of a verification methodology are a common work flow and reusability. Can you imagine if every verification engineer, or designer in some cases, had his or her own way of verifying a block in the design? Each block level testbench would be different from the other, and you could forget about porting anything from the block to the chip level. If you don't have to imagine this scenario because you are currently living it, then migrating to UVM is a definite plus.

Productivity is easily increased with a common methodology because everyone is working from the same template which helps maintain consistency. Files are located in a common directory structure. Components are created with reuse in mind both at the block and chip level.

UVM is supported by all the major simulators and is an Accellera standard. The library is also open source, so it is easy to refer to the source code during development. This book will focus on and use UVM version 1.1.

Planning and Coverage Goals

A separate book could be written about properly creating a verification plan and coverage goals in a metric driven environment. However, whenever you discuss building a verification environment, this subject should be mentioned.

Before you code, write your verification plan. This is a simple statement, but it has huge implications. By writing your verification plan, you have to think through not only what you want to test, but how you are going to test it. Planning ahead of time reduces the number of last minute tests to create. Map your coverage goals to your verification plan. Analyze your coverage results early and often so you can determine where additional tests may be needed. Doing this process continually will help you achieve your goals faster with fewer, more effective tests.

Chapter 2
Testbench Architecture

Testbench Architecture

A typical UVM testbench contains several components. First, let's look at a simple UVM testbench diagram.

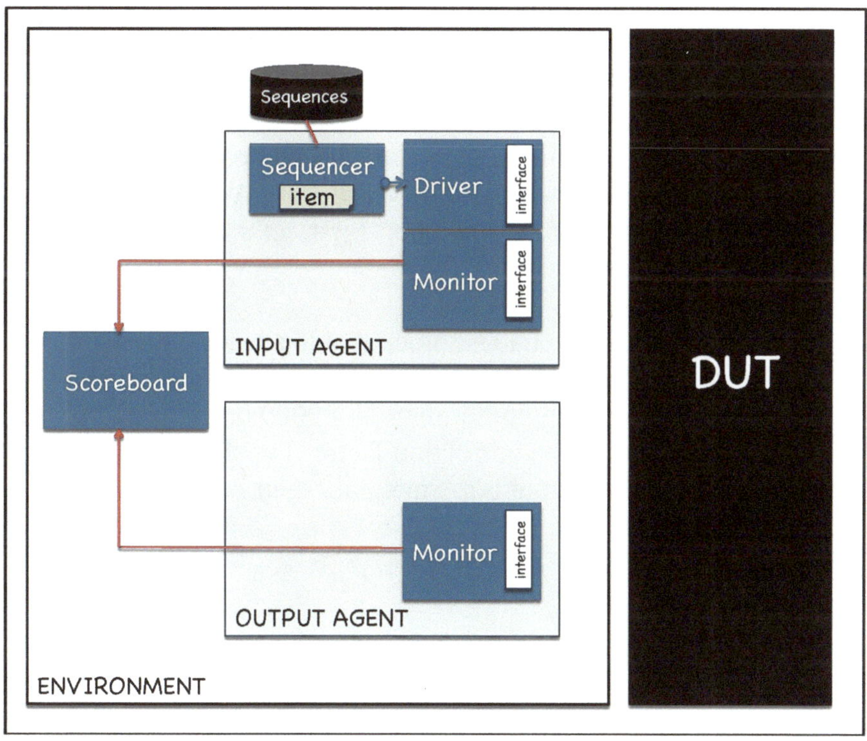

Figure 1: UVM Testbench

There are several components illustrated within this diagram with the most fundamental one being the agent. An agent is a container that holds all the needed components for a particular protocol. A typical UVM agent might contain a driver, monitor, and sequencer. The sequencer controls the flow of sequences that are created either with constrained random or directed data. You can think of a sequence as a packet of data that represents a transaction or protocol. Essentially, a sequencer passes a transaction to the driver. The driver then drives the interface based on the bus protocol with that transaction. The driver can also send a response to the sequencer, if necessary. The monitor captures the transactions based on the protocol and can implement checkers to verify the protocol. All of these pieces, the driver, monitor, etc. are examples of a UVM component. They are reusable pieces of code that together create the protocol you need for your environment.

You will notice that the output agent only contains a monitor. The input agent is called "active" because it is actively driving transactions onto the interface. However, the output agent is "passive." It is just capturing the transactions and passing the information along to the scoreboard for checking. We'll explore how to make an agent active or passive later. First, let's look at an example that better explains this concept.

For our example DUT, we have a block that has two 16 bit data input ports, an enable port, and a 2 bit correction factor port. If the data on the ports is anything other than 16'h0000 or 16'hFFFF, then the data is multiplied by the correction factor if enable is high. If enable is low, then the output ports maintain their previous value.

Figure 2: Design Under Test

Since the output of this block is just the input data transformed and the output has no handshake signals that need to be driven, there is no need for a driver or responder at the output interface.

The following Verilog RTL is provided as a simple example DUT, which will be used to illustrate how to craft the UVM testbench.

```
module pipe( clk,
             rst_n,
             i_cf,
             i_en,
             i_data0,
             i_data1,
             o_data0,
             o_data1
           );

   input          clk;
   input          rst_n;
   input    [1:0] i_cf;
   input          i_en;
   input   [15:0] i_data0;
   input   [15:0] i_data1;

   output  [15:0] o_data0;
   output  [15:0] o_data1;

   wire         clk;
   wire         rst_n;
   wire   [1:0] i_cf;
   wire         i_en;
   wire  [15:0] i_data0;
   wire  [15:0] i_data1;

   reg   [15:0] o_data0;
   reg   [15:0] o_data1;

   reg   [15:0] data_0;
   reg   [15:0] data_1;

   // Store the input data and check to see if it is
   // 16'h0000 or 16'hFFFF
   // if not, multiply by correction factor

   always @(posedge clk) begin
      if(!rst_n) begin
         data_0 <= 16'h0000;
         data_1 <= 16'h0000;
      end
      else begin
         if(i_en) begin
            if((i_data0 == 16'h0000) || (i_data0 == 16'hFFFF)) begin
               data_0 <= i_data0;
            end
            else begin
               data_0 <= i_data0 * i_cf;
            end
            if((i_data1 == 16'h0000) || (i_data1 == 16'hFFFF)) begin
               data_1 <= i_data1;
            end
            else begin
```

```
                data_1 <= i_data1 * i_cf;
            end
         end
      end
   end

   always @(posedge clk) begin
      o_data0 <= data_0;
      o_data1 <= data_1;
   end
endmodule
```

Phases

Before we can build the UVM testbench, there are some basics that need to be covered. The components in the UVM Class library use a set of phases. Phases are used to partition and schedule activities that are common in any simulation. For example, there is a phase to instantiate all the testbench components, a phase where you can initialize elements before simulation, etc. All of these phases are class methods of a UVM component, and they are as follows:

- build_phase
- connect_phase
- end_of_elaboration_phase
- start_of_simulation_phase
- run_phase
- extract_phase, check_phase, and report_phase

build_phase

Although it is not depicted in Figure 1, the top-level is a test and that test builds the testbench. In order to do this, the build_phase is necessary. The build_phase is the first phase in the flow and is the only phase that is called in a top-down manner starting with the test. Its purpose is to create all of the UVM components. Before we look at example code, there are a few key points you need to remember for this phase.

- Every build_phase function should call super.build_phase first so that the build_phase function in the parent class is executed.
- Configuration for a component should be done before a component is created.
- All UVM components should be created in this phase.

Let's look at an example from an AHB agent and discuss the lines of code.

```
function void build_phase(uvm_phase phase);
   super.build_phase(phase);
   if(is_active == UVM_ACTIVE) begin
      sequencer = pipe_sequencer::type_id::create("sequencer",
                  this);
      driver = pipe_driver::type_id::create("driver", this);
   end

   monitor = pipe_monitor::type_id::create("monitor", this);

   `uvm_info(get_full_name( ), "Build stage complete.", UVM_LOW)
endfunction: build_phase
```

Every phase function or task takes an argument of **"phase"** that is of type **"uvm_phase"**. The phase argument contains information regarding the behavior and state of the phase. When you call super.build_phase, you pass phase as an argument there as well. In this build_phase function, we create a driver and sequencer only if this agent is active. The monitor is always created since it is ever-present in active and passive agents. For now, don't worry about "**is_active**" and **"UVM_ACTIVE"**. We'll explore both of those later.

connect_phase

The connect_phase happens directly after the build_phase and does exactly what you would expect. It connects the environment topology. In other words, you have created all of the UVM components in the build_phase and some of those components need to communicate with each other. Those connections are done in this phase. For example, if you need to connect an analysis port from a monitor to the analysis export in a scoreboard, you would do that in the connect_phase. Here's a code example:

```
virtual function void connect_phase(uvm_phase phase);

   //Connect the analysis ports in the monitors to the scoreboard and
coverage

ahb_env.agent.monitor.item_collected_port.connect(scoreboard.ahb_pack
ets_collected.analysis_export);

ahb_env.agent.monitor.item_collected_port.connect(coverage.analysis_e
xport);

   `uvm_info(get_full_name( ), "Connect phase complete.", UVM_LOW)
endfunction: connect_phase
```

In this code snippet, the monitor's connect_phase function calls the analysis port's connect function. The argument to the function is the scoreboard's analysis export. The second connection is from the same analysis port in the AHB monitor to an analysis export in a coverage object. Please don't worry about the terms analysis port and analysis export. The ports and exports are simply a communication mechanism for passing transactions. In the example above, every collected transaction in the monitor is sent to the scoreboard and a coverage object.

end_of_elaboration_phase

This phase is where you should place any post elaboration activity. For example, this is an excellent place to print topology information, which I highly recommend. The resulting printout of the testbench hierarchy in the log file can be a valuable aid in debugging. It also adds documentation in the log file of what was actually created. The UVM Users Guide also gives examples of setting a message verbosity of a component in this phase or setting an address map.

```
virtual function void end_of_elaboration_phase(uvm_phase phase);
   //Print hierarchy information in the end of elaboration phase
```

```
   `uvm_info(get_type_name( ), $sformatf("Printing the test topology
                              :\n%s", this.sprint(printer)), UVM_LOW)

endfunction: end_of_elaboration_phase
```

The printer here is of type uvm_table_printer. It is instantiated in a base test that other tests may inherit from. Here's an example:

```
class base_test extends uvm_test;

   uvm_table_printer printer;

   . . .
   virtual function void build_phase(uvm_phase phase);
      . . .
      printer = new( );
      printer.knobs.depth = 5;
      . . .

      `uvm_info(get_full_name( ), "Build phase complete.", UVM_LOW)
   endfunction: build_phase
endclass: base_test
```

The printer contains knobs that allow you to control what information is printed. In this example, I want the printed topology to only have a recursive depth of 5, i.e., five levels deep into the constructed hierarchy would be printed. If the depth were set to −1, then everything would have been printed. If the concept of control knobs is confusing, picture the volume knob on a radio. By turning this knob, you control how loud the music is played. Now think of an audio soundboard which has dozens of knobs. By tweaking those knobs, you create a configuration of how the audio should sound. Control knobs in the testbench also create a configuration, whether it is test parameters or how deep you want the topology printed.

start_of_simulation_phase

This phase can be used to initialize any components before the simulation starts. You could also print any configuration information for the test you are running for debug reference.

```
virtual function void start_of_simulation_phase(uvm_phase phase);

   `uvm_info(get_type_name( ), {"***** START OF SIMULATION FOR ",
            get_full_name( )}, UVM_LOW)

endfunction: start_of_simulation_phase
```

run_phase

The run_phase is the only phase that is implemented as a task and is time consuming. Any event or thread that needs to occur during the run time execution of your simulation should be in the run_phase. Let's review a simple run_phase task from an AHB driver.

```
virtual task run_phase(uvm_phase phase);
   fork
      reset( );
      get_and_drive( );
      address_phase( );
      data_phase( );
   join
endtask: run_phase
```

In this run_phase task, four other tasks are forked so that they run concurrently. There is a task for reset that drives the signals to zero while reset is active. The get_and_drive task receives the AHB packets which will be sent to the address_phase and data_phase tasks.

Also, within this phase there are a set of twelve run time phases. These phases allow you to further partition when certain events occur such as during reset or shutdown. If you use the run time phases, they should only be used for stimulus partitioning. These phases are not discussed in this book, but more information can be found in the UVM documentation.

extract_phase, check_phase, and report_phase

These phases can be used to extract and collect simulation results or coverage information. A pass or fail status can be determined by the information collected and those results can be reported. These phases provide a mechanism to collate data and perform any final actions.

Usage

None of these functions or the run task must be implemented unless they are being used. How to use some of the phases will become more evident as we build the testbench throughout the book. Also, the source code in this book contains helpful comments about the anticipated role and behavior of these methods.

Macros and Messaging

Macros

The UVM Library has built-in utility and field automation macros that allow objects to be created by the factory and have access to common functions such as copy( ) or clone( ). To understand how the macros work, let's build a simple data packet for the DUT. If we were not using the UVM, a typical data packet may have looked like the following:

```
class data_packet;
   rand bit  [1:0] cf;
   rand bit        enable;
   rand bit [15:0] data_in0;
   rand bit [15:0] data_in1;
   rand bit [15:0] data_out0;
   rand bit [15:0] data_out1;
   rand int        delay;

   constraint timing {delay inside {[0:5]};}
endclass: data_packet
```

Here is a standard SystemVerilog class with several random data members and one constraint block. Now, let's build the UVM version of this.

```
class data_packet extends uvm_sequence_item;
```

The class is now derived from uvm_sequence_item. Every data item that you create must be derived from this base class directly or indirectly. In addition to our data members, we now add the utility and field automation macros:

```
`uvm_object_utils_begin(data_packet)
      `uvm_field_int(cf,        UVM_DEFAULT)
      `uvm_field_int(enable,    UVM_DEFAULT)
      `uvm_field_int(data_in0,  UVM_DEFAULT)
      `uvm_field_int(data_in1,  UVM_DEFAULT)
      `uvm_field_int(data_out0, UVM_DEFAULT)
      `uvm_field_int(data_out1, UVM_DEFAULT)
      `uvm_field_int(delay,     UVM_DEFAULT)
`uvm_object_utils_end
```

The utility macro `uvm_object_utils registers this class with the factory, which we will discuss later, and allows access to the create method which is needed for cloning. After the utility macro are the field automation macros in the form `uvm_field_*(data member, flag). They allow access to the functions copy, compare, pack, unpack, record, print, and sprint. Essentially, these macros work together to implement clone( ), print(), and the other methods that would be tedious and repetitive to code manually. There are macros for the various data types such as integers, enumerations, queues, etc. I recommend reviewing the UVM Reference Guide for an exhaustive list of all of the macros.

The flag indicates what type of automation to enable for that data member. Multiple flag values can be used by using a bit-wise OR. They can also be added using the '+' sign. Here is a table with the flags and their descriptions.

UVM_ALL_ON	Set all operations on (default)
UVM_DEFAULT	Use the default flag settings
UVM_NOCOPY	Do not copy this field.
UVM_NOCOMPARE	Do not compare this field.
UVM_NOPRINT	Do not print this field.
UVM_NOPACK	Do not pack or unpack this field.
UVM_PHYSICAL	Treat as a physical field. Use physical setting in policy class for this field.
UVM_ABSTRACT	Treat as an abstract field. Use the abstract setting in the policy class for this field.
UVM_READONLY	Do not allow setting of this field from the set_*_local methods.
UVM_BIN	Print/record the field in binary (base-2).
UVM_DEC	Print/record the field in decimal (base-10).
UVM_UNSIGNED	Print/record the field in unsigned decimal (base-10).
UVM_OCT	Print/record the field in octal (base-8).
UVM_HEX	Print/record the field in hexadecimal (base-16).
UVM_STRING	Print/record the field in string format.
UVM_TIME	Print/record the field in time format.

If you did not need to use the field macros in your class, then your utility macro would simply be `uvm_object_utils (or `uvm_component_utils for extensions of UVM components) without the _begin and _end.

Here is the entire UVM data packet:

```
class data_packet extends uvm_sequence_item;
   rand bit  [1:0] cf;
   rand bit        enable;
   rand bit [15:0] data_in0;
   rand bit [15:0] data_in1;
   rand bit [15:0] data_out0;
   rand bit [15:0] data_out1;
   rand int        delay;

   constraint timing {delay inside {[0:5]};}

   `uvm_object_utils_begin(data_packet)
      `uvm_field_int(cf,        UVM_DEFAULT)
      `uvm_field_int(enable,    UVM_DEFAULT)
      `uvm_field_int(data_in0,  UVM_DEFAULT)
      `uvm_field_int(data_in1,  UVM_DEFAULT)
      `uvm_field_int(data_out0, UVM_DEFAULT)
      `uvm_field_int(data_out1, UVM_DEFAULT)
      `uvm_field_int(delay,     UVM_DEFAULT)
   `uvm_object_utils_end

  function new(string name = "data_packet");
      super.new(name);
   endfunction: new

endclass: data_packet
```

Adding the macros looks like extra code, but it allows lots of other UVM library code to copy, clone, etc. this object. Now, if you were to cut and paste this code into your favorite editor and then try and compile with your favorite simulation tool, you would receive a whole host of errors. As is, the compiler has no idea of where to find uvm_sequence_item, or `uvm_object_utils, etc. To compile code with UVM objects you must do two things: **import** and **include**. You must import the uvm_pkg and include the uvm_macros. Let's review the updated compile-friendly code:

```
import uvm_pkg::*;
`include "uvm_macros.svh"

class data_packet extends uvm_sequence_item;
  . . .
endclass: data_packet
```

In the actual testbench, you would not want to add these to your individual .sv files. You would want to add them to a package file that included your .sv files for a particular UVM Verification Component, UVC. A UVC is simply a grouping of files that make up a verification component or verification IP. You would then import that package when needed.

Messaging

The testbench needs the ability to report informative messages and debug information. The following macros provide that capability.

- `uvm_info(ID, MSG, VERBOSITY)
- `uvm_warning(ID, MSG)
- `uvm_error(ID, MSG)
- `uvm_fatal(ID, MSG)

With `uvm_info, you not only provide a string ID and string MSG, but you also provide an integer verbosity level. The ID field is simply a tag that you can easily identify in the log. The MSG field is the message you want printed and the verbosity indicates when you want that message printed. If the message's verbosity level is lower or equal to the current verbosity level of that simulation run, then the message is displayed. The verbosity levels, in order of smallest to largest are, UVM_NONE, UVM_LOW, UVM_MEDIUM, UVM_HIGH, UVM_FULL, and UVM_DEBUG.

As an example, let's add a task to the data packet that displays the current values of the data members.

```
virtual function void displayAll( );
   `uvm_info("DP", $sformatf("cf = %0h enable = %0b data_in0 = %0h
                              data_in1 = %0h data_out0 = %0h
                              data_out1 = %0h delay = %0d",
                              cf, enable, data_in0, data_in1,
                              data_out0, data_out1,
                              delay), UVM_LOW)
endfunction: displayAll
```

This message will always print when this task is called because it has the lowest verbosity level, provided the user did not set verbosity to UVM_NONE. If there are messages you only want to print for debug, you would give them a higher verbosity such as **UVM_HIGH** or **UVM_DEBUG**. The default verbosity is **UVM_LOW**. You can change that from the command line with **+UVM_VERBOSITY=UVM_<VERBOSITY_LEVEL>**. It is important to note that macro invocations do not end with a semicolon.

The Factory

When you use the utility macros `uvm_object_utils or `uvm_component_utils, that class is registered with the factory and gives access to the create method. The factory is a class that can create the instances of the objects registered. Creating an instance using the factory is simple.

In a non-factory implementation you would call the constructor of a class to create the instance. For example:

```
dut_driver driver = new("driver", this);
```

When instantiating using the factory, you call the create method:

```
object_name::type_id::create(string name, uvm_component parent);
```

For example:

```
dut_driver driver = dut_driver::type_id::create("driver", this);
```

Using the factory allows you to substitute components within a family when needed without having to create new components or subclasses due to changes in composition. A family of classes is a set of classes derived from a common base class. For example, the driver needs a small modification for one of the blocks you are verifying. Without the factory, you would need to instantiate a new driver in a new agent, which would then require a new environment, etc. If you have to make these types of changes, then the testbench is not resusable.

With the factory, you can override which driver you need without having to change any other component. To override the component you would use either set_type_override_by_type or set_inst_override_by_type. The "type" override replaces all components of that type with the override. The "inst" override replaces targeted components. Here are the prototypes:

```
set_type_override_by_type(orig_type, override_type, bit replace = 1);

set_inst_override_by_type(string inst_path, orig_type,
override_type);
```

To override the driver by type would be the following:

```
set_type_override_by_type(dut_driver::get_type( ),
dut_driver_mod::get_type( ));
```

The third argument, replace, determines whether or not to replace any existing overrides.

To override a specific targeted component would be the following:

```
set_inst_override_by_type("agent1.driver", dut_driver::get_type( ),
dut_driver_mod:get_type( ));
```

Resource and Config DB

uvm_config_db

The uvm_config_db is essentially a library for data sharing. This library or database stores hierarchical configuration values that may be needed across components. The class declaration is as follows:

```
uvm_config_db#(type T = int) extends uvm_resource_db#(T);
```

As an example, a driver or monitor which is deep in the verification component hierarchy has a declaration of a virtual interface. Using the configuration database, you can put the needed interface in the database, and the component that needs it can pull it from the database.

The two most commonly used functions for doing this are get and set, so let's look at some examples:

```
uvm_config_db#(int)::set(this, "*.scoreboard*", "disable_sb", 0);
uvm_config_db#(virtual pipe_if)::set(this, "*.driver", "intf", vif);

uvm_config_db#(int)::get(this, "", "disable_sb", db_val);
uvm_config_db#(virtual pipe_if)::get(this, "", "intf", vif);
```

We will expand upon this example further as the testbench is constructed.

uvm_resource_db

The uvm_config_db is actually built upon the uvm_resource_db. The uvm_resource_db is better suited for cases where a hierarchical context is not needed. The class declaration is as follows:

```
uvm_resource_db#(type T = uvm_object);
```

Here are some examples with the commonly used functions.

```
uvm_resource_db#(string)::set("DUT NAMES", "pipe", dut_name, this);
uvm_resource_db#(string)::read_by_name("DUT NAMES", dut_name, name);
```

As we build the testbench in the following chapters, the use and practicality of these databases will become more evident.

Let's Get Started

The basics have been covered so now it is time to build the testbench. To build a UVM testbench from the ground up, you start with the two most basic elements: the interface and the data transaction. Starting with the transaction, instead of the driver or monitor, allows you to really think about the transactions and sequences that need to be sent. The interface is the element that connects the DUT to the testbench.

SystemVerilog Interfaces

A SystemVerilog interface is a simple construct that allows communication between the design and the testbench. Driving the interface, as opposed to the actual RTL, aids in reuse as well. For example, there are two blocks in your design that have the same inputs. Designer A has a coding style where all of his or her inputs and outputs begin with "i_" or "o_" respectively. Designer B does not follow this style.

By using an interface, the signals in that interface can have a generic, yet descriptive name, regardless of the actual RTL port names.

Here's the interface for the testbench:

```
interface pipe_if(input logic clk, rst_n);
   logic  [1:0] cf;
   logic [15:0] data_in0;
   logic [15:0] data_in1;
   logic [15:0] data_out0;
   logic [15:0] data_out1;
   logic        enable;
endinterface: pipe_if
```

The Data

We have already created the data packet or sequence item. Let's review.

```
class data_packet extends uvm_sequence_item;
```

```
   rand bit  [1:0] cf;
   rand bit        enable;
   rand bit [15:0] data_in0;
   rand bit [15:0] data_in1;
   rand bit [15:0] data_out0;
   rand bit [15:0] data_out1;
   rand int        delay;

   constraint timing {delay inside {[0:5]};}

   `uvm_object_utils_begin(data_packet)
      `uvm_field_int(cf,        UVM_DEFAULT)
      `uvm_field_int(enable,    UVM_DEFAULT)
      `uvm_field_int(data_in0,  UVM_DEFAULT)
      `uvm_field_int(data_in1,  UVM_DEFAULT)
      `uvm_field_int(data_out0, UVM_DEFAULT)
      `uvm_field_int(data_out1, UVM_DEFAULT)
      `uvm_field_int(delay,     UVM_DEFAULT)
   `uvm_object_utils_end

   function new(string name = "data_packet");
      super.new(name);
   endfunction: new

   virtual task displayAll( );
      `uvm_info("DP", $sformatf("cf = %0h enable = %0b data_in0 = %0h
                                data_in1 = %0h data_out0 = %0h
                                data_out1 = %0h delay = %0d", cf,
                                enable, data_in0, data_in1,
                                data_out0, data_out1, delay),
                                UVM_LOW)
   endtask: displayAll

endclass: data_packet
```

So for each transaction, the driver will drive randomized values of cf, data_in0, data_in1, and enable onto the interface. The monitor will capture the output data on the interface for checking with the scoreboard and coverage analysis. Let's start building those components now.

Chapter 3
Drivers

Drivers

The next step in building the UVM testbench is to construct the driver. The driver requests data transactions from the sequencer. Once it has the transactions, it drives the interface signals in adherence to the protocol. First, let's build the driver for the example DUT and then review a driver for a simple AHB-Lite protocol.

Any driver you create will extend uvm_driver. Here is the header for that class:

```
class uvm_driver #(type REQ=uvm_sequence_item, type RSP=REQ) extends
uvm_component;
```

The class is parameterized with a REQ and RSP. REQ is the type of your uvm_sequence_item for sending requests. In this case that would be data_packet, and the driver would receive that item from the sequencer. If the RSP, which is used for returning responses, is different from the REQ, you would specify both types. If they are the same, as in this case, you would only need to specify the type once. If needed, RSP would be a data_packet that is sent back to the sequencer.

Since the DUT is essentially a pipe, input data is transformed into output data, we will call it pipe_driver.

```
class pipe_driver extends uvm_driver #(data_packet);
   virtual pipe_if vif;

   `uvm_component_utils(pipe_driver)

   function new(string name, uvm_component parent);
      super.new(name, parent);
   endfunction: new

   function void build_phase(uvm_phase phase);
      super.build_phase(phase);
      if(!uvm_config_db#(virtual pipe_if)::get(this, "", "in_intf",
                                                vif))
         `uvm_fatal("NOVIF", {"virtual interface must be set for: ",
                              get_full_name( ), ".vif"})
      `uvm_info(get_full_name( ), "Build stage complete.", UVM_LOW)
   endfunction

   virtual task run_phase(uvm_phase phase);
      fork
         reset( );
```

```
        get_and_drive( );
     join
  endtask: run_phase

  virtual task reset( );
     forever begin
        @(negedge vif.rst_n);
        `uvm_info(get_type_name( ), "Resetting signals ", UVM_LOW)
        vif.cf = 2'b0;
        vif.data_in0 = 16'b0;
        vif.data_in1 = 16'b0;
        vif.enable = 1'b0;
     end
  endtask: reset

  virtual task get_and_drive( );
     forever begin
        while(vif.rst_n != 1'b0) begin
           seq_item_port.get_next_item(req);
           drive_packet(req);
           seq_item_port.item_done( );
        end
     end
  endtask: get_and_drive

  virtual task drive_packet(data_packet pkt);
     vif.enable = 1'b0;
     repeat(pkt.delay) @(posedge vif.clk);
     vif.enable = pkt.enable;
     vif.cf = pkt.cf;
     vif.data_in0 = pkt.data_in0;
     vif.data_in1 = pkt.data_in1;
     @(posedge vif.clk);
     vif.enable = 1'b0;
  endtask

endclass:pipe_driver
```

The pipe_driver class is parameterized with data_packet which we defined earlier. As a reminder, that packet contains the following:

```
  rand bit  [1:0] cf;
  rand bit        enable;
  rand bit [15:0] data_in0;
  rand bit [15:0] data_in1;
  rand bit [15:0] data_out0;
  rand bit [15:0] data_out1;
  rand int        delay;
```

With the exception of delay, the variables here represent the inputs and outputs of the pipe. The variable delay is used to add delay between transactions. Let's examine the first section of code.

```
  virtual pipe_if vif;

  `uvm_component_utils(pipe_driver)
```

```
function new(string name, uvm_component parent);
   super.new(name, parent);
endfunction: new

function void build_phase(uvm_phase phase);
   super.build_phase(phase);
   if(!uvm_config_db#(virtual pipe_if)::get(this, "", "in_intf",
                       vif))
      `uvm_fatal("NOVIF", {"virtual interface must be set for: ",
                            get_full_name( ), ".vif"})
   `uvm_info(get_full_name( ), "Build stage complete.", UVM_LOW)
endfunction
```

First, the driver declares the virtual interface, pipe_if, that it will be driving. It then calls the utility macro `uvm_component_utils to enable automation and register this class with the factory. Next is a typical UVM constructor with a string and uvm_component arguments.

The most interesting part of this section is the build_phase where the uvm_config_db is used to obtain the virtual interface. Note, I have encapsulated the call to uvm_config_db in an if statement and inverted it. If the call is unsuccessful, then the `uvm_fatal macro will trigger and end the simulation with a "NOVIF" error. Is this extra step necessary? No. Will it save you hours of debug? Yes.

The next section is the run_phase. Remember, the run_phase is the only phase that consumes time and all test execution occurs during this phase.

```
virtual task run_phase(uvm_phase phase);
    fork
       reset( );
       get_and_drive( );
    join
endtask: run_phase
```

This task calls two other tasks, reset and get_and_drive, in a fork/join block for parallel execution. The reset task will reset the signals to 0 based on the status of the active low reset signal. The get_and_drive task will initiate communication with the sequencer to get the transaction.

```
virtual task reset( );
   forever begin
      @(negedge vif.rst_n);
```

```
            `uvm_info(get_type_name( ), "Resetting signals ", UVM_LOW)
            vif.cf = 2'b0;
            vif.data_in0 = 15'b0;
            vif.data_in1 = 15'b0;
            vif.enable = 1'b0;
         end
      endtask: reset

      virtual task get_and_drive( );
         forever begin
            while(vif.rst_n != 1'b0) begin
               seq_item_port.get_next_item(req);
               drive_packet(req);
               seq_item_port.item_done( );
            end
         end
      endtask: get_and_drive
```

Notice that in both tasks, the heart of the execution is encased in a forever block. This is necessary so that these threads execute continuously until the end of the simulation. When you are writing your own testbench and accidentally code an infinite forever loop, remember to check that you have a blocking statement in your loop or a statement that advances time. In the reset task, it blocks until the negedge of vif.rst_n.

Let's examine the get_and_drive task in more detail. The first statement of interest is the following:

```
seq_item_port.get_next_item(req);
```

The seq_item_port is used to get items from the sequencer. It calls the task get_next_item which blocks until an item is available. The argument here is "req" which is a data member of the base class uvm_driver, and its type is REQ. REQ stands for request transaction type. The packet is then sent to the drive_packet task. Once that task has completed, the seq_item_port then calls the item_done task which indicates to the sequencer that the sequence is done.

Finally, the drive_packet task drives the transaction values onto the interface after a randomized delay.

```
   virtual task drive_packet(data_packet pkt);
      vif.enable = 1'b0;
      repeat(pkt.delay) @(posedge vif.clk);
      vif.enable = pkt.enable;
```

```
        vif.cf = pkt.cf;
        vif.data_in0 = pkt.data_in0;
        vif.data_in1 = pkt.data_in1;
        @(posedge vif.clk);
        vif.enable = 1'b0;
    endtask
```

Now, let's review a slightly different driver implementation with the AHB-Lite protocol. When developing a driver for this protocol, you must keep in mind that it is pipelined. If you used the previous example, you would send the packet to a drive task. That task would then drive the address during the address phase and then on the next clock, drive the data during the data phase. However, since the next address phase occurs during the current data phase, that transaction would be missed because the sequence is not done. So you would need to implement the driver in such a way that the address phase communicates with the data phase without blocking the next transaction. There are several ways to do this, but for this example we will use mailboxes.

In addition to declaring a virtual interface, the driver also declares an address_box and data_box as type mailbox.

```
mailbox #(ahb_transfer) address_box = new(1);
mailbox #(ahb_transfer) data_box = new(1);
```

The run_phase task would have a task for the address phase and the data phase.

```
    virtual task run_phase(uvm_phase phase);
        fork
            reset( );
            get_and_drive( );
            address_phase( );
            data_phase( );
        join
    endtask: run_phase
```

The get_and_drive task for this driver has to handle the packet in a smarter way than our simple driver.

```
    virtual task get_and_drive( );
        ahb_transfer pkt;
        forever begin
            while(vif.hresetn != 1'b0) begin
```

```
            seq_item_port.get_next_item(req);
            $cast(pkt, req.clone( ));
            address_box.put(pkt);
            seq_item_port.item_done( );
         end
      end
   endtask: get_and_drive
```

First it declares a packet, “pkt”, of the ahb_transfer type which is the transaction type for this driver. The transaction is received via the seq_item_port as usual. Now, instead of being driven immediately, it is cloned into the new packet and put into the mailbox, address_box. The packet is cloned so that any new req that is received does not overwrite the original. Once the packet is put into the mailbox, the sequence is done and waits for the next transaction. Notice that the packet was not sent to another task, but simply put into the mailbox to be picked up by the address_phase.

```
  virtual task address_phase( );
      ahb_transfer addr_pkt;
      forever begin
         vif.htrans <= IDLE;
         address_box.get(addr_pkt);
         while(vif.hready == NOT_READY) begin
            @(posedge vif.hclk);
         end

         vif.haddr <= addr_pkt.haddr;
         vif.hburst <= addr_pkt.hburst;
         vif.hmastlock <= addr_pkt.hmastlock;
         vif.hprot <= addr_pkt.hprot;
         vif.hsize <= addr_pkt.hsize;
         vif.hwrite <= addr_pkt.hwrite;
         vif.htrans <= addr_pkt.htrans;
         vif.hsel <= 1'b1;  /* Only one slave */


         @(posedge vif.hclk);
         data_box.put(addr_pkt);
         vif.haddr <= 'h0;
      end
   endtask: address_phase
```

The address phase waits for a packet in its mailbox. Once it is received, it drives the interface signals adhering to the address phase portion of the protocol. Once done, the packet is placed in the data_box for the data_phase.

```
   virtual task data_phase( );
      ahb_transfer data_pkt;
```

```
    forever begin
       data_box.get(data_pkt);

       if(data_pkt.hwrite == WRITE) begin
           vif.hwdata <= data_pkt.hwdata;
           `uvm_info(get_type_name( ), $sformatf("WRITING DATA: %0h
                     at %0t", data_pkt.hwdata, $time), UVM_LOW)

           while(vif.hready  == NOT_READY) begin
              @(posedge vif.hclk);
           end
        end

        if(data_pkt.hwrite == READ) begin
           @(posedge vif.hclk);
           while(vif.hready == NOT_READY) begin
              @(posedge vif.hclk);
           end
           data_pkt.hrdata = vif.hrdata;
          `uvm_info(get_type_name( ), $sformatf("READING DATA: %0h
                     at %0t", vif.hrdata, $time), UVM_LOW)
        end
        seq_item_port.put(data_pkt);
    end
 endtask: data_phase
```

Just as the address_phase did, the data_phase waits for its mailbox to receive a packet. Once it does, it drives the data phase portion of the protocol. If it is a read transaction, the read data is stored in the packet and returned to the sequencer.

These examples illustrate two different ways to handle the transaction once it is received from the sequencer. Now let's review the sequencer and how to generate sequences.

Chapter 4
Sequencers and Sequences

Sequencers

I once gave a UVM demo where I was explaining to the engineers that the driver received the transaction through the sequencer. One asked, “Well, what does the sequencer look like?” I showed him the following code.

```
class ahb_sequencer extends uvm_sequencer #(ahb_transfer);

   `uvm_sequencer_utils(ahb_sequencer)

   function new(string name, uvm_component parent);
      super.new(name, parent);
   endfunction: new

endclass: ahb_sequencer
```

“That’s it?” he replied. “That’s it!” was my response. The fact that you can write six lines of code yet have a powerful implementation is the beauty of UVM. Having the sequencer mechanism prepared for you in the library allows you to build the testbench faster and more effectively. Let’s briefly examine the connection between the sequencer and the driver.

The sequencer’s seq_item_export is connected to the driver’s seq_item_port, and it defines the functions used to retrieve packets as was discussed in the last chapter.

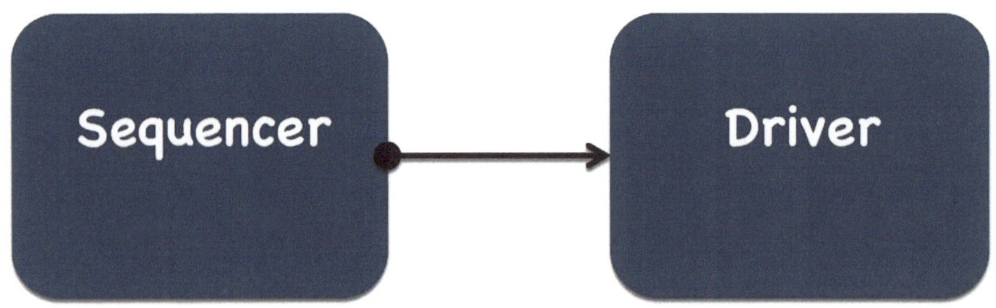

Figure 3: Sequencer and Driver Connection

```
driver.seq_item_port.connect(sequencer.seq_item_export);
```

This connection is made in the agent, which we will examine later. For completeness, here is the sequencer for the example pipe DUT.

```
class pipe_sequencer extends uvm_sequencer #(data_packet);

   `uvm_sequencer_utils(pipe_sequencer)

   function new(string name, uvm_component parent);
      super.new(name, parent);
   endfunction: new

endclass: pipe_sequencer
```

Essentially, the job of the sequencer is to control the flow of sequences to the driver. In the next section, we will explore a few ways to generate sequences.

Sequences

Now is the perfect time to review graphically the components we have built thus far.

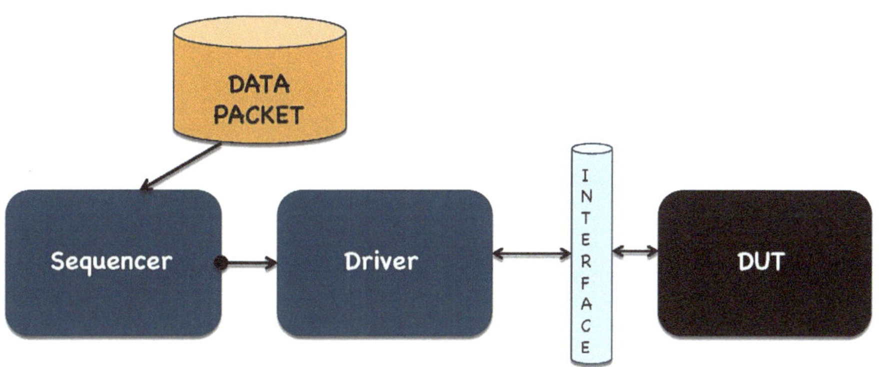

Figure 4: Sequences

We have our data packet, the interface, a driver, and a sequencer. Now we need to learn how to create sequences using the data packet. A sequence specifies one or more sequence items to be sent to the driver; in other words, one or more packets to be driven. Typically, sequence classes are derived from uvm_sequence and parameterized with your uvm_sequence_item type. For example:

```
class random_sequence extends uvm_sequence #(data_packet);
```

Your sequence class will have a task, body( ), that starts the sequence. Let's look at the first example with the **`uvm_do** macro.

```
class random_sequence extends uvm_sequence #(data_packet);
   `uvm_object_utils(random_sequence)

   function new(string name = "random_sequence");
      super.new(name);
   endfunction:new

   virtual task body( );
      `uvm_do(req);
   endtask: body
endclass: random_sequence
```

As mentioned with the driver, **req** is a member of uvm_sequence that is now parameterized as data_packet. The macro, **`uvm_do**, creates the transacation if it is not already created, randomizes it, and sends it to the sequencer. Of course, this only sends one transaction. You can easily use a for loop to send several random packets.

```
class many_random_sequence extends uvm_sequence #(data_packet);
   rand int loop;

   constraint limit {loop inside {[5:10]};}
   `uvm_object_utils(many_random_sequence)

   function new(string name = "many_random_sequence");
      super.new(name);
   endfunction:new

   virtual task body( );
      for(int i = 0; i < loop; i++) begin
        `uvm_do(req);
      end
   endtask: body

endclass: many_random_sequence
```

A companion macro that you will use often is **`uvm_do_with**. As with **`uvm_do**, it creates your transaction, however, it now randomizes it with the inline constraints that you provide. Finally, it sends it to the sequencer. For example, let's constrain data_in0 to be 0.

```
class data0_sequence extends uvm_sequence #(data_packet);
   `uvm_object_utils(data0_sequence)

   function new(string name = "data0_sequence");
      super.new(name);
   endfunction:new

   virtual task body( );
      `uvm_do_with(req, {req.data_in0 == 16'h0;})
   endtask: body

endclass: data0_sequence
```

Another way to start a sequence item is to use **start_item/finish_item**. The function, **start_item,** initiates the start of the sequence, but the item has to be created first. You then randomize and constrain that item before calling **finish_item**. In the following example, I have created the item before the **start_item** function.

```
class data1_sequence extends uvm_sequence #(data_packet);
   `uvm_object_utils(data1_sequence)

   function new(string name = "data1_sequence");
      super.new(name);
   endfunction:new

   virtual task body( );
      req = data_packet::type_id::create("req");
      start_item(req);
      assert(req.randomize( ) with {data_in1 == 'hffff;});
      finish_item(req);
   endtask: body

endclass: data1_sequence
```

One question you may have is how to connect the sequence to the sequencer. One way this is done is by calling the sequence's start function with the sequencer as an argument. We will explore how to do that in the chapter on Tests. For now, let's continue to build the testbench by creating the monitor.

Chapter 5
Monitors

Monitors

The monitor is a passive component in the testbench that serves two functions: it verifies the protocol and collects transactions for checking and coverage. While developing the monitor, if you find yourself duplicating the code in the driver, stop. The monitor should not be a duplication of the driver but a complementary element that understands the protocol so it can sample the transactions at the right time.

When the transaction is sampled or collected, it is then written to an analysis port. Although not a completely accurate analogy, since it implies storage, but I like to think of the analysis port as a dropbox. You drop the collected transaction in the box where it can then be picked up by whatever component needs it, such as a scoreboard or coverage object. Let's review the monitor for the pipe example and then examine it section by section.

```
class pipe_monitor extends uvm_monitor;
   virtual pipe_if vif;
   string monitor_intf;
   int num_pkts;

   uvm_analysis_port #(data_packet) item_collected_port;
   data_packet data_collected;
   data_packet data_clone;

   `uvm_component_utils(pipe_monitor)

   function new(string name, uvm_component parent);
      super.new(name, parent);
   endfunction: new

   function void build_phase(uvm_phase phase);
      super.build_phase(phase);
      if(!uvm_config_db#(string)::get(this, "", "monitor_intf",
                                        monitor_intf))
         `uvm_fatal("NOSTRING", {"Need interface name for: ",
                    get_full_name( ), ".monitor_intf"})

      `uvm_info(get_type_name( ), $sformatf("INTERFACE USED = %0s",
                                             monitor_intf), UVM_LOW)
      if(!uvm_config_db#(virtual pipe_if)::get(this, "",
                              monitor_intf, vif))
         `uvm_fatal("NOVIF", {"virtual interface must be set for: ",
                     get_full_name( ), ".vif"})

      item_collected_port = new("item_collected_port", this);
```

```
      data_collected =
      data_packet::type_id::create("data_collected");

      data_clone = data_packet::type_id::create("data_clone");

      `uvm_info(get_full_name( ), "Build stage complete.", UVM_LOW)
   endfunction: build_phase

   virtual task run_phase(uvm_phase phase);
      collect_data( );
   endtask: run_phase

   virtual task collect_data( );
      forever begin
         wait(vif.enable)
         data_collected.cf = vif.cf;
         data_collected.data_in0 = vif.data_in0;
         data_collected.data_in1 = vif.data_in1;
         repeat(3) @(posedge vif.clk);
         data_collected.data_out0 = vif.data_out0;
         data_collected.data_out1 = vif.data_out1;
         $cast(data_clone, data_collected.clone( ));
         item_collected_port.write(data_clone);
         num_pkts++;
      end
   endtask: collect_data

   virtual function void report_phase(uvm_phase phase);
      `uvm_info(get_type_name( ), $sformatf("REPORT: COLLECTED
                PACKETS = %0d", num_pkts), UVM_LOW)
   endfunction: report_phase

endclass: pipe_monitor
```

Let's examine the first section of code.

```
   virtual pipe_if vif;
   string monitor_intf;
   int num_pkts;

   uvm_analysis_port #(data_packet) item_collected_port;
   data_packet data_collected;
   data_packet data_clone;

   `uvm_component_utils(pipe_monitor)

   function new(string name, uvm_component parent);
      super.new(name, parent);
   endfunction: new
```

I have declared a virtual interface, vif, a string, monitor_intf, and an int, num_pkts. The monitor collects the data from the signals on this virtual interface. The int, num_pkts, is simply used to keep track of how many transactions are collected. The string, monitor_intf, is used in conjunction with the configuration database. I will expand on this statement further when we examine the build_phase.

Next I have declared item_collected_port, which is an analysis port. The analysis port is parameterized to the data_packet. Once you have collected the data for one transaction, you call the analysis port's write function to deposit that data. The data_packet, data_collected, is the packet used to store the data from the interface. Once complete, data_collected is cloned to data_clone and that cloned packet is written to the analysis port.

Cloning the packet before you write it to the analysis port is crucial. When you pass the transaction object as an argument to the write function, UVM does this using a pass-by-reference, and therefore it does not create a new object. If you were to send the original data packet collected, it would be altered as you collect the next transaction. As an anecdotal example, the first time I did this the packet I was collecting contained four queues. I wrote the collected data to the analysis port without cloning. Since I did not want the queues to continue to grow with each transaction, I deleted the data in the queues to start the next transaction with a clean slate. Unwittingly, I had also deleted the data in the queues that I was trying to use in my scoreboard.

With this approach, I am making the assumption that the cloned data packet will not be changed by a subscribing object. If it is possible that the object could be changed in the scoreboard, coverage model, or any other subscriber, you want to clone the object in the subscriber.

The last part of this section is the utility macro, **`uvm_component_utils**, which registers pipe_monitor with the factory and the standard constructor for a component. The next section is the build_phase.

```
function void build_phase(uvm_phase phase);
   super.build_phase(phase);
   if(!uvm_config_db#(string)::get(this, "", "monitor_intf",
                                      monitor_intf))
      `uvm_fatal("NOSTRING", {"Need interface name for: ",
                get_full_name( ), ".monitor_intf"})
```

```
    `uvm_info(get_type_name( ), $sformatf("INTERFACE USED = %0s",
                                  monitor_intf), UVM_LOW)
    if(!uvm_config_db#(virtual pipe_if)::get(this, "",
                         monitor_intf, vif))
      `uvm_fatal("NOVIF", {"virtual interface must be set for: ",
                    get_full_name( ), ".vif"})

    item_collected_port = new("item_collected_port", this);
    data_collected =
    data_packet::type_id::create("data_collected");
    data_clone = data_packet::type_id::create("data_clone");

    `uvm_info(get_full_name( ), "Build stage complete.", UVM_LOW)
  endfunction: build_phase
```

Two things of note are happening in the build_phase. First, I am using the configuration database to retrieve the virtual interface. Before discussing the syntax of the code, let's review a diagram and examine why it is necessary.

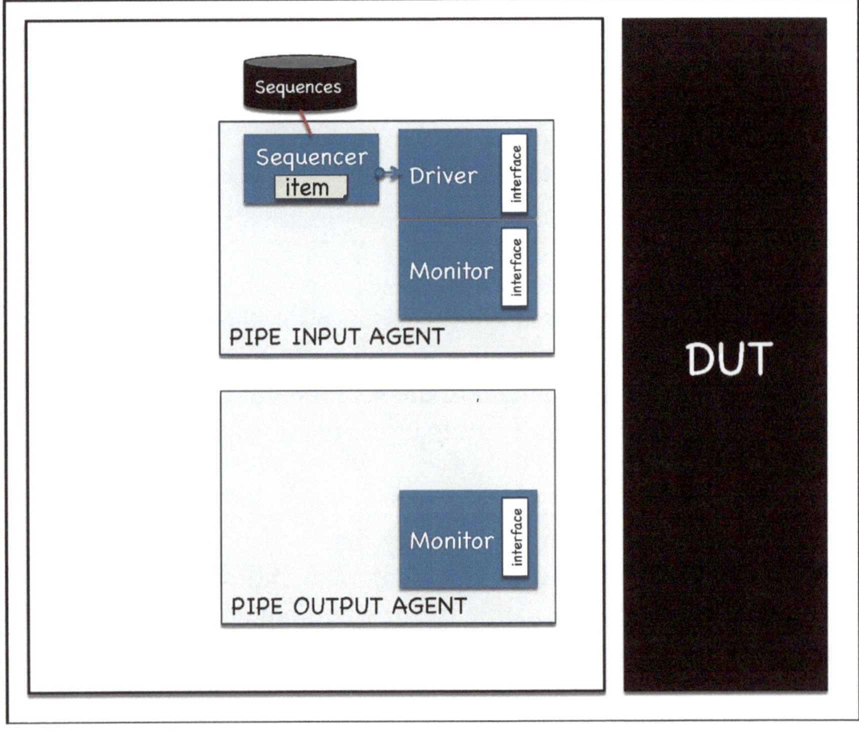

Figure 5: Input and Output Agents

As shown in the diagram, the monitor is being reused in the input agent and the output agent. The output agent is the same as the input agent, except that it is set as "passive" and only creates the monitor. We will review agents in the next chapter. There are two instances of the interface, so two configuration database entries are needed to ensure that the monitor retrieves the correct one. These interfaces are declared and instantiated in the top level.

```
pipe_if ivif(.clk(clk), .rst_n(rst_n));
pipe_if ovif(.clk(clk), .rst_n(rst_n));
```

To store the interfaces in the database, the set function is used:

```
uvm_config_db#(virtual pipe_if)::set(uvm_root::get( ) , "*" ,
               "in_intf", ivif);
uvm_config_db#(virtual pipe_if)::set(uvm_root::get( ) , "*" ,
               "out_intf", ovif);
```

To break down what the set function is doing, "**uvm_root::get( )**", is getting the top level and setting that as the context. Using the wildcard is making this entry available globally. Finally, you are giving the interface instance, **ivif**, the name or label "**in_intf**" in the database. The output interface is labeled "**out_intf**".

Using the wildcard in this case is acceptable given the small nature of the testbench. However, in a larger design you want to limit how often you have a database entry globally available. Making it globally available could have a performance impact as well as affect debug. For example, if you were trying to debug a configuration entry and you have 200 that are all globally available, printing them out and trying to find the one line that you need would be difficult. It is better to limit the scope of availability if possible. A better method would be the following:

```
uvm_config_db#(virtual pipe_if)::set(uvm_root::get( ) , "*.agent.*" ,
               "in_intf", ivif);
uvm_config_db#(virtual pipe_if)::set(uvm_root::get( ) , "*.monitor" ,
               "out_intf", ovif);
```

With this change, the input interface is now only available to everything under agent since both the monitor and driver need access to it. The output interface is limited to only monitors.

As mentioned, the monitor needs to retrieve the appropriate interface, but it needs to remain generic for reuse. To solve this problem, we declare a string in the monitor that will get the label of the interface from the database and use that to retrieve the interface. That string was mentioned earlier and it is "**monitor_intf**". Environments will be discussed in the next chapter, but that is where the database is used to associate the interface with **monitor_intf**.

```
uvm_config_db#(string)::set(this, "penv_in.agent.monitor",
                  "monitor_intf", "in_intf");
uvm_config_db#(string)::set(this, "penv_out.agent.monitor",
                  "monitor_intf", "out_intf");
```

With this call to the set function, we have limited the availability to either the monitor in the input environment or the output environment. For the input, "**in_intf**" is the name or label for the input interface. It is given label "**monitor_intf**". Note, this **monitor_intf** is a label in the database and not the **monitor_intf** that is the string in the monitor. You can use any name you like here. I prefer to use the same name to mentally associate the two to each other.

The build phase of the monitor is the final step in this process.

```
if(!uvm_config_db#(string)::get(this, "", "monitor_intf",
                    monitor_intf))
   `uvm_fatal("NOSTRING", {"Need interface name for: ",
                get_full_name( ), ".monitor_intf"})

`uvm_info(get_type_name( ), $sformatf("INTERFACE USED = %0s",
                                     monitor_intf), UVM_LOW)

if(!uvm_config_db#(virtual pipe_if)::get(this, "", monitor_intf,
                    vif))
   `uvm_fatal("NOVIF", {"virtual interface must be set for: ",
                get_full_name( ), ".vif"})
```

The first call to get retrieves the value stored in "**monitor_intf**" in the database for either the input or the output. It then assigns that string value to the string **monitor_intf**. For debug purposes, it prints out the value of **monitor_intf,** and then the final get call assigns the name of the interface which is stored in the string **monitor_intf** to the virtual interface instance, **vif**.

The final part of the build_phase is creating the objects the monitor will be using.

```
item_collected_port = new("item_collected_port", this);
data_collected = data_packet::type_id::create("data_collected");
data_clone = data_packet::type_id::create("data_clone");
```

The uvm_analysis_port is not registered with the factory, so to create it you call the constructor as usual. I have used the factory to create data_collected and data_clone. In this case, using the factory is not necessary since these objects will not need to be overridden, but it is good practice to use the factory whenever possible.

The run_phase simply calls the task **collect_data( )** where the transaction collection is done.

```
virtual task collect_data( )
   forever begin
      wait(vif.enable)
      data_collected.cf = vif.cf;
      data_collected.data_in0 = vif.data_in0;
      data_collected.data_in1 = vif.data_in1;
      repeat(3) @(posedge vif.clk);
      data_collected.data_out0 = vif.data_out0;
      data_collected.data_out1 = vif.data_out1;
      $cast(data_clone, data_collected.clone( ));
      item_collected_port.write(data_clone);
      num_pkts++;
   end
endtask: collect_data
```

In this forever loop, it first waits for the enable signal of the interface. Once it is high, it collects the input values on the interface and assigns them to their corresponding values in data_collected. After three clock cycles, the output data is valid and so that is collected. Again, once the data is collected, it's cloned and assigned to data_clone with $cast. The packet, data_clone, is finally written to the analysis_port with a call to the analysis port's write function.

The final portion of the monitor is the report_phase.

```
virtual function void report_phase(uvm_phase phase);
   `uvm_info(get_type_name( ), $sformatf("REPORT: COLLECTED PACKETS =
%0d", num_pkts), UVM_LOW)
endfunction: report_phase
```

Each time a packet is written to the analysis port, num_pkts is incremented. The number of packets collected is then printed in the report_phase after the run_phase has completed. Remember, this monitor is for both the input and the output, so seeing the number of packets collected and comparing the two in the scoreboard is a good method to check that a packet was not lost.

The three components of an agent, driver, sequencer, and monitor, have now been built. In the next chapter, we will examine how to construct the agent and an environment.

Chapter 6
Agents and Environments

Agents and Environments

Agents

As mentioned, an agent is a container that instantiates the driver, monitor, and sequencer. Agents can be either active or passive. In active mode, all three components are created, while in passive mode only the monitor is created. Since we have already built the necessary components, creating the agent is simply a matter of instantiating them. Let's review the code.

```
class pipe_agent extends uvm_agent;
   protected uvm_active_passive_enum is_active = UVM_ACTIVE;

   pipe_sequencer sequencer;
   pipe_driver    driver;
   pipe_monitor   monitor;

   `uvm_component_utils_begin(pipe_agent)
      `uvm_field_enum(uvm_active_passive_enum, is_active, UVM_ALL_ON)
   `uvm_component_utils_end

   function new(string name, uvm_component parent);
      super.new(name, parent);
   endfunction

   function void build_phase(uvm_phase phase);
      super.build_phase(phase);
      if(is_active == UVM_ACTIVE) begin
         sequencer = pipe_sequencer::type_id::create("sequencer",
                     this);
         driver = pipe_driver::type_id::create("driver", this);
      end

      monitor = pipe_monitor::type_id::create("monitor", this);

      `uvm_info(get_full_name( ), "Build stage complete.", UVM_LOW)
   endfunction: build_phase

   function void connect_phase(uvm_phase phase);
      if(is_active == UVM_ACTIVE)
         driver.seq_item_port.connect(sequencer.seq_item_export);
      `uvm_info(get_full_name( ), "Connect stage complete.", UVM_LOW)
   endfunction: connect_phase
endclass: pipe_agent
```

The first declaration in the agent is the variable "is_active" which is of the type "uvm_active_passive_enum". This enumerated type can be found in the UVM library and has two possible values:

UVM_PASSIVE and UVM_ACTIVE. I have assigned it UVM_ACTIVE, but this can be changed to UVM_PASSIVE when needed via the configuration database. Notice that I have included "is_active" in the field automation list with the `uvm_field_enum list. By doing this, when I instantiate this agent using the factory "is_active" will be given the value in the configuration database and that will override the default. This concept will be illustrated momentarily when we discuss the environment.

The build_phase function implementation is quite simple. For an agent configured to be active, it creates the sequencer, driver, and then the monitor. If the agent is configured to be passive, then it only creates the monitor. In the connect_phase, we once again see the use of the connect function. Here, the driver's communication port is connected to the sequencer's communication export.

Once again, this is why UVM is so powerful. In one line of code it appears that I have connected my driver and sequencer so that sequence items can be passed and executed. I used the word appears because the library does the hard part for you. There is an underlying transaction level modeling, TLM, communication machine that handles the handshake between the driver and the sequencer. Now, it is good to read the code in the UVM library to understand how this works, but to get started all you need to know is that you use the connect function in one line of code.

Now, the agent that we just created needs to be instantiated. This instantiation is typically done in an environment.

Environments

An environment is another container. This particular container holds agents and other environments. It also contains other objects that are needed for the simulation such as a scoreboard, register model, memory models, coverage objects, etc. It's called an environment because it contains the components that are necessary to create an effective verification environment. Let's review a couple of diagrams. In the first, I use the environment simply to instantiate an agent.

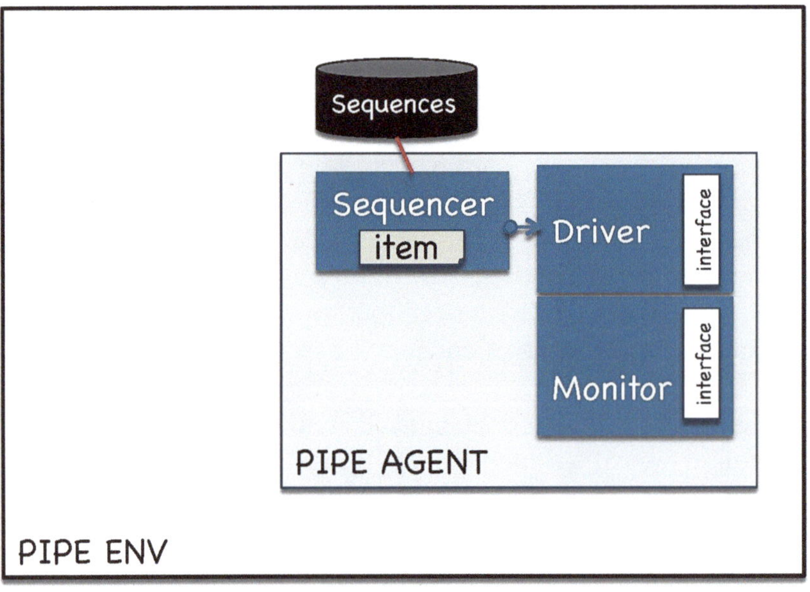

Figure 6: Pipe Environment

The code for this environment is as simple as the diagram looks. In the build_phase, I simply use the factory to create the agent object.

```
class pipe_env extends uvm_env;
```

```
   pipe_agent agent;

   `uvm_component_utils(pipe_env)

   function new(string name, uvm_component parent);
      super.new(name, parent);
   endfunction

   function void build_phase(uvm_phase phase);
      super.build_phase(phase);

      agent = pipe_agent::type_id::create("agent", this);

      `uvm_info(get_full_name( ), "Build stage complete.", UVM_LOW)
   endfunction: build_phase

endclass: pipe_env
```

In the second diagram, I have created an environment that instantiations my pipe environment twice. One instantiation is for the input and the other is for the output. It will also instantiate my scoreboard.

Figure 7: DUT Environment

Let's take a look at the code for this environment.

```
class dut_env extends uvm_env;

   pipe_env        penv_in;
   pipe_env        penv_out;
   pipe_scoreboard sb;

   `uvm_component_utils(dut_env)

   function new(string name, uvm_component parent);
      super.new(name, parent);
   endfunction

   function void build_phase(uvm_phase phase);
      super.build_phase(phase);

      uvm_config_db#(int)::set(this, "penv_in.agent", "is_active",
                      UVM_ACTIVE);
      uvm_config_db#(int)::set(this, "penv_out.agent", "is_active",
                      UVM_PASSIVE);

      uvm_config_db#(string)::set(this, "penv_in.agent.monitor",
                      "monitor_intf", "in_intf");
      uvm_config_db#(string)::set(this, "penv_out.agent.monitor",
                      "monitor_intf", "out_intf");

      penv_in = pipe_env::type_id::create("penv_in", this);
      penv_out = pipe_env::type_id::create("penv_out", this);
      sb = pipe_scoreboard::type_id::create("sb", this);

      `uvm_info(get_full_name( ), "Build stage complete.", UVM_LOW)
   endfunction: build_phase

   function void connect_phase(uvm_phase phase);
      penv_in.agent.monitor.item_collected_port.connect(
      sb.input_packets_collected.analysis_export);

      penv_out.agent.monitor.item_collected_port.connect(
      sb.output_packets_collected.analysis_export);

      `uvm_info(get_full_name( ), "Connect phase complete.", UVM_LOW)
   endfunction: connect_phase

endclass: dut_env
```

As expected, I have created input and output environments for the pipe and the scoreboard. However, before they are created in the build_phase, I use the configuration database to set some values.

```
uvm_config_db#(int)::set(this, "penv_in.agent", "is_active",
                UVM_ACTIVE);
uvm_config_db#(int)::set(this, "penv_out.agent", "is_active",
                UVM_PASSIVE);
```

In the lines above, I am setting the input agent to be active and the output agent to be passive. It is important that I set these values in the database before I actually create the objects. Now, when the agent objects are created, the items in the field automation list, such as **"is_active"** will get the values stored in the database. What if you do this in reverse order? If that's the case, then you must use the static get function of the configuration database to retrieve the values.

In the connect_phase, as we have seen before, I'm connecting the input and output monitors to the scoreboard for checking. One key element missing from this environment is a coverage object. In the next chapter, we will explore how to develop a functional coverage object in UVM, but you now have the tools to instantiate and connect that coverage object in the environment.

Chapter 7
Scoreboards and Coverage

Scoreboards

The basic function of the scoreboard is to check the correctness of the output data of the design under test. The scoreboard you create should derive from uvm_scoreboard; however, there is no current functionality of the uvm_scoreboard. To illustrate this point, review the file in your installation. You will find it in /uvm-1.1/src/comps/uvm_scoreboard.svh.

You may be wondering why it's important to mention if it doesn't have any functionality at the moment. The important aspect is how the scoreboard retrieves its data for comparison. To better understand how this is is done, let's further examine analysis ports and analysis exports.

Analysis Ports

If you recall from the monitor, there was an analysis port used to broadcast the collected data. Let's revisit a portion of that code.

```
uvm_analysis_port #(data_packet) item_collected_port;
. . .
item_collected_port.write(data_clone);
```

An analysis port is a TLM communication port that has a write function. In the monitor, we collected the data, cloned it, and used the port's write function to broadcast this data to any subscribers. The subscriber in this case is the scoreboard. It will pick up this broadcasted data via its analysis export.

Analysis Exports

The analysis export of the subscriber or scoreboard must implement the write function. One way to do this is by using a uvm_tlm_analysis_fifo. The benefit of using the fifo is that it has an analysis export, implements the needed write function, and has an unbounded queue for storing transactions. Let's review the declarations for this fifo.

```
uvm_tlm_analysis_fifo #(data_packet) input_packets_collected;
uvm_tlm_analysis_fifo #(data_packet) output_packets_collected;
```

The first line declares an analysis fifo called input_packets_collected that is parameterized as a data packet. This fifo will be used to collect transactions from the input monitor. Conversely, output_packets_collected is for the output data. Let's look at the entire scoreboard implementation and discuss it section by section.

```
class pipe_scoreboard extends uvm_scoreboard;
   uvm_tlm_analysis_fifo #(data_packet) input_packets_collected;
   uvm_tlm_analysis_fifo #(data_packet) output_packets_collected;

   data_packet input_packet;
   data_packet output_packet;

   `uvm_component_utils(pipe_scoreboard)

   function new(string name, uvm_component parent);
      super.new(name, parent);
   endfunction: new

   function void build_phase(uvm_phase phase);
      super.build_phase(phase);

      input_packets_collected = new("input_packets_collected", this);
      output_packets_collected = new("output_packets_collected",
                                    this);

      input_packet = data_packet::type_id::create("input_packet");
      output_packet = data_packet::type_id::create("output_packet");

      `uvm_info(get_full_name( ), "Build stage complete.", UVM_LOW)
   endfunction: build_phase

   virtual task run_phase(uvm_phase phase);
      watcher( );
   endtask: run_phase

   virtual task watcher( );
      forever begin
         input_packets_collected.get(input_packet);
         output_packets_collected.get(output_packet);
         compare_data( );
      end
   endtask: watcher

   virtual task compare_data( );
      bit [15:0] exp_data0;
      bit [15:0] exp_data1;

      if((input_packet.data_in0 == 16'h0000) ||
```

```
         (input_packet.data_in0 == 16'hffff))
         exp_data0 = input_packet.data_in0;
      else
         exp_data0 = input_packet.data_in0 * input_packet.cf;

      if((input_packet.data_in1 == 16'h0000) ||
         (input_packet.data_in1 == 16'hffff))
         exp_data1 = input_packet.data_in1;
      else
         exp_data1 = input_packet.data_in1 * input_packet.cf;

      if(exp_data0 != output_packet.data_out0)
         `uvm_error(get_type_name( ), $sformatf("Actual output data
          %0h does not match expected %0h", output_packet.data_out0,
          exp_data0))

      if(exp_data1 != output_packet.data_out1)
         `uvm_error(get_type_name( ), $sformatf("Actual output data
         %0h does not match expected %0h", output_packet.data_out1,
         exp_data1))

   endtask:compare_data

endclass: pipe_scoreboard
```

The first portion of this code should look familiar. I have declared the analysis fifos and an input and output packet. I have the constructor and the build phase to create the objects. Please note that for the uvm_analysis_tlm_fifos, you instantiate them using their constructor rather than the factory.

The run phase simply calls a task named watcher which is inside a forever loop. The watcher task first waits for the input by using the blocking get function of the uvm_analysis_tlm_fifo. The output of the get function is the input_packet. Once it has the input_packet, it blocks until it has the output_packet. It then calls a compare function which compares the output data to the expected output data based on the algorithm of the DUT.

As with the example DUT, this is a simple scoreboard but it should illustrate an example of communication between a port and export. Let's review a second example with a coverage object.

Coverage

The coverage object will extend the uvm_subscriber class and be parameterized with the data_packet. Since the object is of type uvm_subscriber, it has an analyis_export and must implement the write function. Let's review a simple example.

```
class pipe_coverage extends uvm_subscriber #(data_packet);

   data_packet pkt;
   int count;

   `uvm_component_utils(pipe_coverage)

   covergroup cg;
      option.per_instance = 1;
      cov_cf:    coverpoint pkt.cf;
      cov_en:    coverpoint pkt.enable;
      cov_ino:   coverpoint pkt.data_in0;
      cov_in1:   coverpoint pkt.data_in1;
      cov_out0:  coverpoint pkt.data_out0;
      cov_out1:  coverpoint pkt.data_out1;
      cov_del:   coverpoint pkt.delay;
   endgroup: cg

   function new(string name, uvm_component parent);
      super.new(name, parent);
      cg = new( );
   endfunction: new

   function void write(data_packet t);
      pkt = t;
      count++;
      cg.sample( );
   endfunction: write

   virtual function void extract_phase(uvm_phase phase);
      `uvm_info(get_type_name( ), $sformatf("Number of coverage
       packets collected = %0d", count), UVM_LOW)

      `uvm_info(get_type_name( ), $sformatf("Current coverage  = %f",
       cg.get_coverage( )), UVM_LOW)
   endfunction: extract_phase
endclass: pipe_coverage
```

The pipe_coverage class has a typical covergroup with coverpoints of the elements in the data packet. The write function receives an instance of data_packet from the monitor through the analysis export. It assigns that instance to the member packet of the class, increments the received count, and calls the covergroup's sample function.

During the extract phase, which occurs after the run phase has completed, I retrieve how many packets were sampled and the coverage data. The get_coverage function will give you the percentage covered.

You must instantiate this coverage class in your environment and use the connect function to enable communication between the analysis port and export.

If you have planned in the beginning and written stimulus for the coverage goals, then this number should be fairly high. You can use your simulator's coverage analysis tool to examine the holes you have missed and grow your test library. In the next chapter, we will construct a test library using the sequences we developed earlier.

Chapter 8
Tests

Stimulus

One of the most often asked UVM questions is "How do I run a test?" To begin to answer that question, let's first look at a diagram to understand the topology.

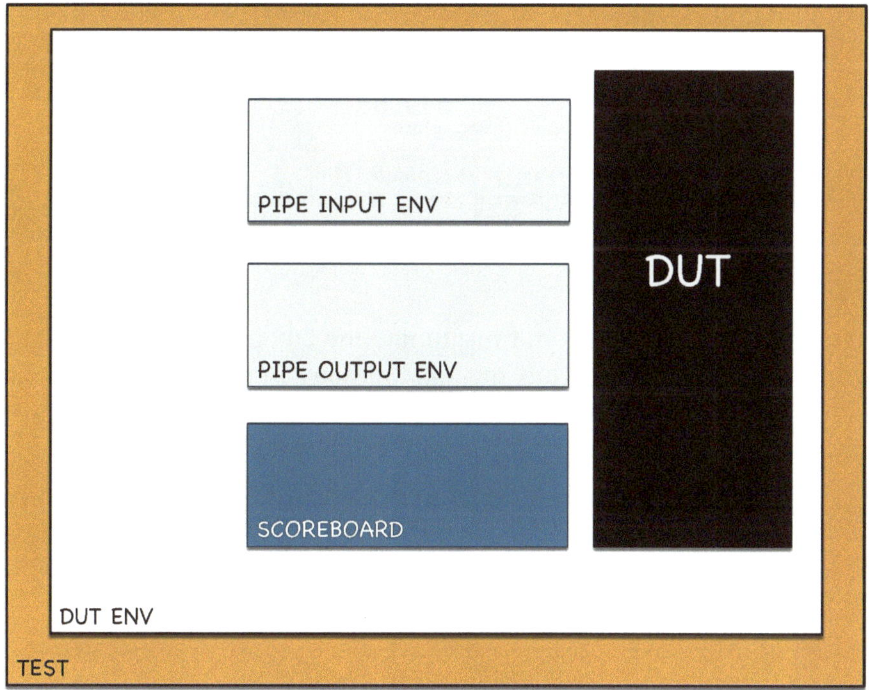

Figure 8: Test Topology

As you can see from the diagram, a test instantiates the environment. Each test is a class that derives from uvm_test. A test library is simply a collection of tests that stimulate the DUT. When building a test library, I prefer to start with a base test from which other tests can derive. This base test would include elements that are required by all tests, such as the environment. Let's review an example.

```
class base_test extends uvm_test;
  `uvm_component_utils(base_test)

  dut_env env;
  uvm_table_printer printer;
```

```
   function new(string name, uvm_component parent);
      super.new(name, parent);
   endfunction: new

   function void build_phase(uvm_phase phase);
      super.build_phase(phase);
      env = dut_env::type_id::create("env", this);
      printer = new( );
      printer.knobs.depth = 5;
   endfunction:build_phase

   virtual function void end_of_elaboration_phase(uvm_phase phase);
      `uvm_info(get_type_name( ), $sformatf("Printing the test
                                  topology :\n%s",
                                  this.sprint(printer)), UVM_DEBUG)
   endfunction: end_of_elaboration_phase

   virtual task run_phase(uvm_phase phase);
      phase.phase_done.set_drain_time(this, 1500);
   endtask: run_phase

endclass: base_test
```

In my build_phase function, I instantiate the env and a uvm_table_printer that prints the test topology in the end_of_elaboration_phase. Printing out the topology can be great for debugging your hierarchy. I have set the verbosity of this to UVM_DEBUG so that I can easily print it only when I need to do debug. If we were to print out the topology of testbench, we would get the following:

```
UVM_INFO test_lib.sv(21) @ 0: uvm_test_top [many_random_test] Printing the test topology :
-------------------------------------------------------------------------------
Name                          Type                          Size  Value
-------------------------------------------------------------------------------
uvm_test_top                  many_random_test              -     @4806
  env                         dut_env                       -     @292
    penv_in                   pipe_env                      -     @4888
      agent                   pipe_agent                    -     @5155
        driver                pipe_driver                   -     @5363
          rsp_port            uvm_analysis_port             -     @6639
          sqr_pull_port       uvm_seq_item_pull_port        -     @6566
        monitor               pipe_monitor                  -     @6413
          item_collected_port uvm_analysis_port             -     @6794
        sequencer             pipe_sequencer                -     @5368
          rsp_export          uvm_analysis_export           -     @5537
          seq_item_export     uvm_seq_item_pull_imp         -     @6421
          arbitration_queue   array                         0     -
          lock_queue          array                         0     -
          num_last_reqs       integral                      32    'd1
          num_last_rsps       integral                      32    'd1
        is_active             uvm_active_passive_enum       1     UVM_ACTIVE
    penv_out                  pipe_env                      -     @5021
      agent                   pipe_agent                    -     @6855
        monitor               pipe_monitor                  -     @6632
          item_collected_port uvm_analysis_port             -     @7046
        is_active             uvm_active_passive_enum       1     UVM_PASSIVE
    pipe_cov                  pipe_coverage                 -     @5088
      analysis_imp            uvm_analysis_imp              -     @5296
    sb                        pipe_scoreboard               -     @4937
      input_packets_collected uvm_tlm_analysis_fifo #(T)    -     @7141
        analysis_export       uvm_analysis_imp              -     @7518
        get_ap                uvm_analysis_port             -     @7442
        get_peek_export       uvm_get_peek_imp              -     @7292
        put_ap                uvm_analysis_port             -     @7367
        put_export            uvm_put_imp                   -     @7219
      output_packets_collected uvm_tlm_analysis_fifo #(T)   -     @7511
        analysis_export       uvm_analysis_imp              -     @7961
        get_ap                uvm_analysis_port             -     @7885
        get_peek_export       uvm_get_peek_imp              -     @7735
        put_ap                uvm_analysis_port             -     @7810
        put_export            uvm_put_imp                   -     @7662
-------------------------------------------------------------------------------
```

Figure 9: Topology

At the top of the topology is a test called many_random_test. From the diagram, you can see each level of instantiation.

Finally, in the run_phase of the base test, we set a drain time. This is adding simulation time to allow all elements to complete after the final objection has been lowered. We will examine objections momentarily. Let's take a look at the first test to derive from the base test.

```
class random_test extends base_test;
   `uvm_component_utils(random_test)

   function new(string name, uvm_component parent);
      super.new(name, parent);
   endfunction: new
```

```
   function void build_phase(uvm_phase phase);
      super.build_phase(phase);
   endfunction: build_phase

   virtual task run_phase(uvm_phase phase);
      random_sequence seq;

      super.run_phase(phase);
      phase.raise_objection(this);
      seq = random_sequence::type_id::create("seq");
      seq.start(env.penv_in.agent.sequencer);
      phase.drop_objection(this);
   endtask: run_phase
endclass: random_test
```

In the run_phase of this test, we first declare a handle to a sequence object called random_sequence. The class random_sequence simply created a random data_packet that was sent to the sequencer. For review, refer to the chapter on Sequences or view the complete code listing in the Appendix.

After calling super.run_phase, we raise an objection with the raise_objection method. The objection mechanism is used to communicate when it is safe to end a phase. By raising the objection, it is an indication that the phase is still in progress. After the objection is raised, the sequence is created using the factory, and then it is launched with the start method. Notice that the argument for the start method is the sequencer for this particular sequence. After the sequence has completed, the drop_objection method is called indicating it is now safe to end this phase.

You may have noticed that we deviated from the norm here by creating our sequence object in the run phase and not the build phase. Sequences do not have phases and are not elements that need to persist throughout the simulation. Although you can create them in the build phase, it is more appropriate to do so in the run phase so that they can be created and destroyed as needed.

For other test examples, including the many_random_test shown in the topology diagram, please see the Appendix.

Starting a Test

You now know how to create a test. To actually start the test, a task called run_test is called from the initial block in your top level module. This task either takes the test name as a string argument, or more commonly, you specify the test name on the command line with UVM_TESTNAME. For example:

+UVM_TESTNAME=random_test

Let's review the top level for our testbench example.

```
module top;
   import uvm_pkg::*;
   import pipe_pkg::*;

   bit clk;
   bit rst_n;

   pipe_if ivif(.clk(clk), .rst_n(rst_n));
   pipe_if ovif(.clk(clk), .rst_n(rst_n));


   pipe pipe_top(.clk(clk),
                 .rst_n(rst_n),
                 .i_cf(ivif.cf),
                 .i_en(ivif.enable),
                 .i_data0(ivif.data_in0),
                 .i_data1(ivif.data_in1),
                 .o_data0(ovif.data_out0),
                 .o_data1(ovif.data_out1)
                );

   always #5 clk = ~clk;

   initial begin
       #5 rst_n = 1'b0;
      #25 rst_n = 1'b1;
   end

   assign ovif.enable = ivif.enable;

   initial begin
      uvm_config_db#(virtual pipe_if)::set(uvm_root::get( ) ,
                     "*.agent.*" , "in_intf", ivif);
      uvm_config_db#(virtual pipe_if)::set(uvm_root::get( ) ,
                     "*.monitor" , "out_intf", ovif);

      run_test( );
   end

endmodule
```

In module top, I have imported the uvm package and the pipe package which contains all the class declarations needed for simulation. I've instantiated the input and output interfaces as well as the DUT. In the initial block, the configuration database is used to store the interfaces. As a review, the input interface is made available to both the driver and monitor since they are instantiated by the agent. The output interface is only available to the monitor. Finally we have the call to run_test which creates the test based on the name and then the components in the various build_phase methods top down.

You now have a fully functional UVM testbench!

Chapter 9
Introduction to the Register Package

Introduction to the Register Package

With the exception of the simple design in the book, most designs have a set of registers. UVM now has a register package that allows you to control the registers in your DUT. It also serves as a model of your DUT registers and a scoreboard for checking. There are plenty of tools available that will generate the register model. I highly recommend taking advantage of one of those tools since creating one manually, especially if there are a lot of registers, can be quite tedious and error prone. In this chapter, we will take a look at the elements of a register model, data prediction, and checking.

Register Components

The register is the most fundamental component of the model and it is of type uvm_reg. Its data members are register fields which are of type uvm_reg_field. Let's look at an example register class for a control register.

```
class regs_control_reg extends uvm_reg;
   rand uvm_reg_field enable;
   rand uvm_reg_field mode_select;

   `uvm_object_utils(regs_control_reg)

   function new(string name = "regs_control_reg");
      super.new(name, 32, UVM_NO_COVERAGE);
   endfunction: new

   virtual function void build( );
      this.enable = uvm_reg_field::type_id::create("enable");
      this.enable.configure(this, 16, 0, "RW", 0, 1'h0, 1, 0, 0);

      this.mode_select =
      uvm_reg_field::type_id::create("mode_select");
      this.mode_select.configure(this, 16, 16, "RW", 0,
                                 1'h0, 1, 0, 0);
   endfunction: build
endclass: regs_control_reg
```

The class is called regs_control_reg and it has two fields: enable and mode_select. You will notice in the constructor, that the call to super.new has three arguments. The second argument is the bit length of the register and the third is the coverage model. The register package has several predefined coverage model identifiers. Here I have indicated that I do not want coverage. I could have used UVM_CVR_ALL if I wanted to use all of the coverage models.

The build function of this class is where the register fields are instantiated using the factory and configured. Here is the declaration for the configure method.

```
function void configure(uvm_reg      parent,
                        int unsigned size,
                        int unsigned lsb_pos,
                        string       access,
                        bit volatile,
                        uvm_reg_data_t reset,
                        bit has_reset,
                        bit is_rand,
                        bit individually_accessible)
```

So the enable field is 16 bits long and starts at position 0 in the register. It is read writable, non-volatile, and has a reset value of 0. The configure field is the same with the exception that it starts at position 16 within the register. As another example, let's review a status register whose field is read only and volatile.

```
class regs_status_reg extends uvm_reg;
   rand uvm_reg_field status;

   `uvm_object_utils(regs_status_reg)

   function new(string name = "regs_status_reg");
      super.new(name, 32, UVM_NO_COVERAGE);
   endfunction: new

   virtual function void build( );
      this.status = uvm_reg_field::type_id::create("status");
      this.status.configure(this, 32, 0, "RO", 1, 1'h0, 1, 0, 0);
   endfunction: build
endclass: regs_status_reg
```

The status register only has one field, aptly named status. In the configure method, the access is set to read only and the volatile field, directly to the right of access, is set to 1. If a register is volatile, it can be updated by the design without the register model being updated because there is no bus transaction. In the current UVM releases, 1.1 and 1.1a, the default for a register read is to check the value against the model's value even if it is volatile. In this sense, the volatility bit has no meaning. However, this may change in future releases where if the field is volatile, checking would be disabled by default.

The next level of hierarchy in the register model is the register block. The register block is a class extended from uvm_reg_block and it instantiates the registers. A register block can also instantiate other register blocks.

```
class dut_regs extends uvm_reg_block;
   rand regs_control_reg control_reg;
   rand regs_status_reg  status_reg;

   `uvm_object_utils(dut_regs)

   function new(string name = "dut_regs");
      super.new(name, UVM_NO_COVERAGE);
   endfunction: new

   virtual function void build( );
      this.default_map = create_map("dut_reg_map", 0, 4,
      UVM_LITTLE_ENDIAN, 0);

      this.control_reg =
      regs_control_reg::type_id::create("control_reg");

      this.control_reg.build( );
      this.control_reg.configure(this, null, "");
      this.default_map.add_reg(this.control_reg,
      `UVM_REG_ADDR_WIDTH'h04, "RW", 0);

      this.status_reg =
      regs_status_reg::type_id::create("status_reg");

      this.status_reg.build( );
      this.status_reg.configure(this, null, "");
      this.default_map.add_reg(this.status_reg,
      `UVM_REG_ADDR_WIDTH'h08, "RO", 0);

      lock_model( );
   endfunction: build

endclass: dut_regs
```

As with the registers, the register block implements a mandatory build function. The build function is called after the register block is created in the environment. The first item in the build function is to create the map for this block by calling the create_map function. The map locates the offset addresses of the registers. Here is the method declaration.

```
virtual function uvm_reg_map create_map(string name,
                                         uvm_reg_addr_t base_addr,
                                         int unsigned n_bytes,
                                         uvm_endianness_e endian,
                                         bit byte_addressing = 1)
```

Here the "dut_reg_map" uses a little endian type, has a base address of 0, and each address is 4 bytes in width. After creating the map, the registers are created using the factory and their respective build methods are called. After the build method is called, the uvm_reg configure method is called. As you can see from the argument list in the example, this is different from the configure method of uvm_reg_field. Here is the declaration.

```
function void configure (uvm_reg_block       blk_parent,
                         uvm_reg_file        regfile_parent = null,
                         string       hdl_path = "")
```

The uvm_reg configure method accepts the parent register block, the parent register file if there is one, and the HDL path for backdoor access. Backdoor means accessing the design register directly via its hierarchical path. After the configure method, the register is added to the map with the add_reg method.

```
virtual function void add_reg (uvm_reg       rg,
                               uvm_reg_addr_t offset,
                               string rights = "RW",
                               bit unmapped = 0,
                               uvm_reg_frontdoor frontdoor = null)
```

The control_reg is added with an offset of 'h04 and is read-writable; whereas the status_reg has an offset of 'h08 and is read only. The define `UVM_REG_ADDR_WIDTH by default is 64 but that can be easily changed to 32 or whatever is desired. Finally, we lock the model so that no further changes can be made.

Register Adapter

Now that the register model is built, you would instantiate it using the factory inside your env class. However, before you can do any bus transaction there needs to be a way to convert an item from the register model to a bus type item and vice versa. To do this you need to implement a register adapter. Let's review a diagram.

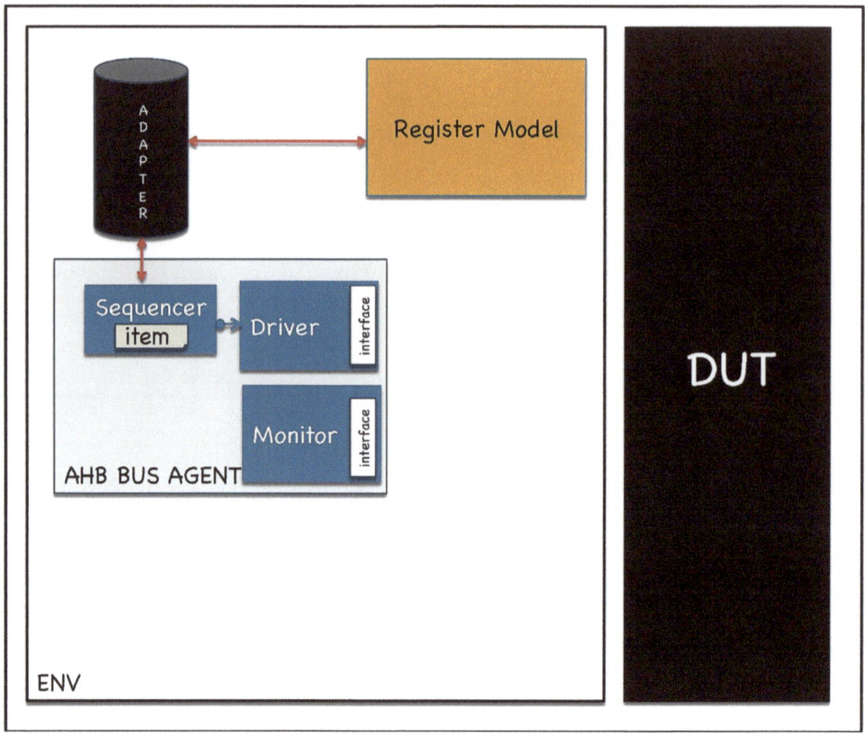

Figure 10: Register Adapter

The register adapter extends the uvm_reg_adapter class. It must implement two functions: reg2bus and bus2reg. Let's review an example with an AHB bus transaction.

```
class reg2ahb_adapter extends uvm_reg_adapter;

   `uvm_object_utils(reg2ahb_adapter)

   function new(string name = "reg2ahb_adapter");
      super.new(name);
      provides_responses = 1;
   endfunction: new

   virtual function uvm_sequence_item reg2bus(const ref
   uvm_reg_bus_op rw);

      ahb_transfer ahb_pkt =
```

```
      ahb_transfer::type_id::create("ahb_pkt");

      assert(ahb_pkt.randomize( ));

      ahb_pkt.hwrite = (rw.kind == UVM_READ) ? READ : WRITE;
      ahb_pkt.haddr = rw.addr;

      if(rw.kind == UVM_WRITE) begin
         ahb_pkt.hwdata = rw.data;
      end
      else begin
         ahb_pkt.hrdata = rw.data;
      end
      return ahb_pkt;
   endfunction: reg2bus

   virtual function void bus2reg(uvm_sequence_item bus_item,
   ref uvm_reg_bus_op rw);

      ahb_transfer ahb_pkt;
      if(!$cast(ahb_pkt, bus_item)) begin
         `uvm_fatal("NOT AHB TYPE", "Provided bus_item is not of the
          correct type")
         return;
      end
      rw.kind = ahb_pkt.hwrite ? UVM_WRITE : UVM_READ;
      rw.addr = ahb_pkt.haddr;
      if(ahb_pkt.hwrite == UVM_WRITE) begin
         rw.data = ahb_pkt.hwdata;
      end
      else begin
         rw.data = ahb_pkt.hrdata;
      end
      rw.status = UVM_IS_OK;
   endfunction: bus2reg

endclass: reg2ahb_adapter
```

In the constructor, I have set provides_responses to 1. This variable indicates to the adapter that the AHB driver will provide a response item. The reg2bus function takes a bus operation from the register model as an argument. An AHB packet is created and randomized. The address, data, and transaction type, read or write, are assigned to the appropriate signals in the packet, and that AHB packet is returned.

The bus2reg function takes both a bus operation from the register model and a bus_item as arguments. The bus_item is compared to an ahb_pkt to confirm it is the correct type. If so, then the signals from the bus_item are assigned to the register model.

Usage and Prediction

With the adapter in place, the model can be used to write and read the registers inside the design from a sequence that extends uvm_reg_sequence. The two most common methods used are write and read. Here are their declarations.

```
virtual task write(output uvm_status_e       status,
                   input  uvm_reg_data_t value,
                   input  uvm_path_e path = UVM_DEFAULT_PATH,
                   input  uvm_reg_map         map =  null,
                   input  uvm_sequence_base parent = null,
                   input  int          prior =  -1,
                   input  uvm_object extension =  null,
                   input  string fname =  "",
                   input  int          lineno =  0)

virtual task read(output uvm_status_e status,
                  output uvm_reg_data_t value,
                  input  uvm_path_e path =  UVM_DEFAULT_PATH,
                  input  uvm_reg_map map = null,
                  input  uvm_sequence_base parent =  null,
                  input  int  prior =  -1,
                  input  uvm_object extension =  null,
                  input  string fname =  "",
                  input  int lineno =  0)
```

For example, if the instance of our dut_regs class is regmodel, a write to the control register would be as follows:

```
regmodel.control_reg.write(status, 32'h5555_aaaa, .parent(this));
```

A status value is returned indicating whether the write transaction was successful. If the return value of status is UVM_NOT_OK, then you would want to flag a `uvm_error. The second argument is the write data for this transaction. Not only is it written to the actual register in the design, but the register model is updated with this value as well. Finally, it is given a pointer to the calling class as its parent. Doing a read is quite similar. Here is an example of a read of the status register.

```
regmodel.status_reg.read(status, read_data, .parent(this));
```

When doing a read, the second argument is the read data that is returned from the register. The value in the register model will be updated with the value that is read on the bus during this transaction. This brings us to the topics of checking and prediction.

If I were to write to the control register and then do a read, the model can check that the read data is the same as the data stored in the register model from the previous write. To do this in UVM 1.1, instead of using the read function to read the register, you would use the mirror function and set its check field to UVM_CHECK.

```
virtual task mirror(output uvm_status_e status,
                    input  uvm_check_e check = UVM_NO_CHECK,
                    input  uvm_path_e path = UVM_DEFAULT_PATH,
                    input  uvm_reg_map map =  null,
                    input  uvm_sequence_base parent =  null,
                    input  int prior =  -1,
                    input  uvm_object extension      =  null,
                    input  string fname =  "",
                    input  int lineno =  0)
```

If you are using UVM 1.1a or later, then there is a more intuitive way of checking. There is a function, set_check_on_read, that when passed a value of 1 will set checking for all registers in its map. For example:

```
regmodel.default_map.set_check_on_read(1);
```

This call to set_check_on_read should be done in the connect_phase function of your env. Once this is done, you can use the read function instead of mirror, and the read values will be checked.

I made the statement earlier that when a write or read transaction is issued the register model would be updated with the respective values. That statement is true, but it is not the default behavior. To understand how to enable this, we must examine the three types of prediction: implicit, explicit, and passive.

Implicit Prediction

Implicit prediction is the mode I described earlier where you have a register model, you issue writes and reads, and the model is updated. However, for the model to be updated, you must call the set_auto_predict function with a value of 1. For example:

```
regmodel.default_map.set_auto_predict(1);
```

This too should be done in the connect_phase of the env class. When using this mode of prediction, the model "thinks" it knows what the values of the actual registers are because this mode of prediction does not pick up transactions to the design registers that did not originate from the register model. Implicit prediction is great for sequences because you are just doing transactions, but it is not good for a scoreboard that needs to retrieve accurate register information to validate functionality.

Explicit Prediction

With explicit prediction, the register model is also integrated with the bus monitor. With this integration, the register model can be updated even when it does not initiate the transaction because it updated via the monitor. In this case, set_auto_predict would be passed a value of 0 and a uvm_reg_predictor needs to be instantiated. The predictor is the component that observes the transaction from the monitor and then calls the predict function in the register model to do the update. A diagram will illustrate the relationship between these components.

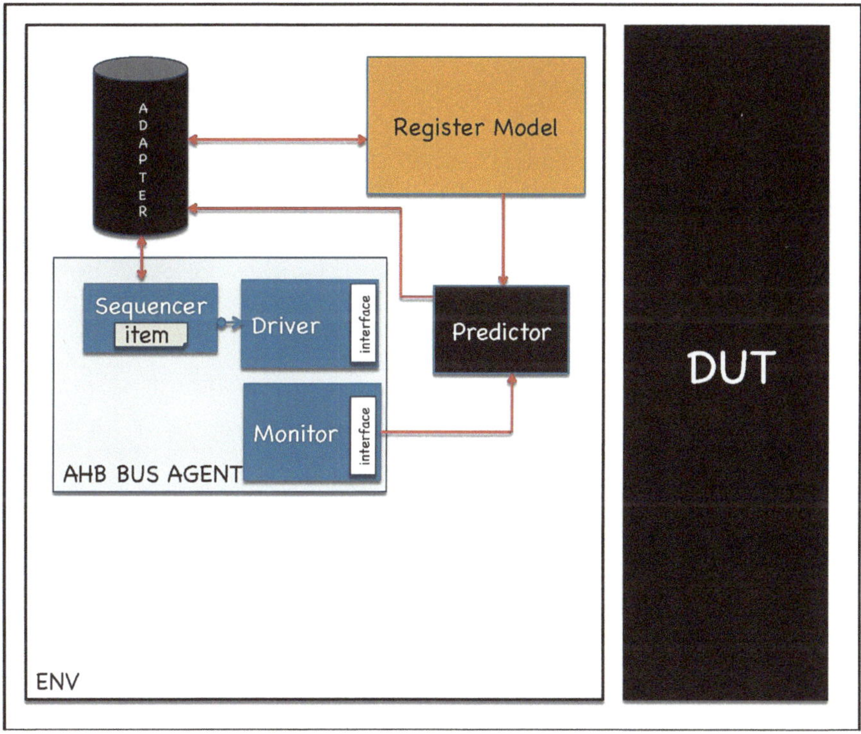

Figure 11: Register Predictor

In your env you would instantiate the uvm_reg_predictor parameterized with your bus sequence item type. In the connect_phase you would do the following:

- Set the predictor map to the register model map.
- Set the predictor adapter to the register adapter.
- Connect the predictor to the monitor.

This mode of prediction can be used to issue bus transactions, but is also excellent for checking since it is up to date with the values in the RTL.

Passive Prediction

Passive prediction is exactly like explicit prediction with one exception; it is not connected to a sequencer. With no sequencer, read and write transactions cannot be issued. This mode of prediction is used for checking only.

Let's review a code example that you would add to the existing dut_env class.

```
class dut_env extends uvm_env;

   ...
   ahb_env ahb_bus_env;
   dut_regs regmodel;
   dut_regs sb_regmodel;
   reg2ahb_adapter stim_adapter;
   reg2ahb_adapter sb_adapter;
   uvm_reg_predictor#(ahb_transfer) mpredictor;

   `uvm_component_utils(dut_env)

   function new(string name, uvm_component parent);
      super.new(name, parent);
   endfunction

   virtual function void build_phase(uvm_phase phase);
      super.build_phase(phase);

      ...
      ahb_bus_env = ahb_env::type_id::create("ahb_bus_env", this);

      regmodel = dut_regs::type_id::create("regmodel", this);
      sb_regmodel = dut_regs::type_id::create("sb_regmodel", this);

      mpredictor =
      uvm_reg_predictor#(ahb_transfer)::type_id::create("mpredictor",
      this);

      ...
   endfunction: build_phase

   virtual function void connect_phase(uvm_phase phase);
      ...

      //Connect the Register Models
      if(regmodel.get_parent( ) == null) begin
         stim_adapter =
         reg2ahb_adapter::type_id::create("stim_adapter",,
         get_full_name( ));

         regmodel.default_map.set_sequencer(
         ahb_bus_env.agent.sequencer, stim_adapter);

         regmodel.default_map.set_auto_predict(1);
```

```
      end

      if(sb_regmodel.get_parent( ) == null) begin
         sb_adapter = reg2ahb_adapter::type_id::create("sb_adapter",,
         get_full_name( ));

         sb_regmodel.default_map.set_sequencer(
         ahb_bus_env.agent.sequencer, sb_adapter);

         mpredictor.map = sb_regmodel.default_map;
         mpredictor.adapter = sb_adapter;
         sb_regmodel.default_map.set_auto_predict(0);
         sb_regmodel.default_map.set_check_on_read(1);

         ahb_bus_env.agent.monitor.item_collected_port.connect(
         mpredictor.bus_in);
      end
      ...

   endfunction: connect_phase

endclass: dut_env
```

I have declared two instances of the register model: regmodel and sb_regmodel. The first one, regmodel, is for sequences only and will use implicit prediction. The second one, sb_regmodel, will use explicit prediction so that it can perform accurate checking and that data could be used in a scoreboard. In the connect_phase, I create the adapter for regmodel and associate the adapter and the sequencer to this model with the set_sequencer method of the map. I then set the auto predict for implicit prediction. I do the same for the sb_regmodel, but the predictor is also configured. Here I set the auto predict for explicit prediction and enable checking via the read function with the set_check_on_read call. It would also be perfectly valid to use passive prediction here since the sb_regmodel will not be issuing transactions. To make it passive prediction, the set_sequencer call would not be made.

This chapter was meant to serve as an introduction of the usage of the register model. I highly encourage you to take a look at the User's Guide for more information as you begin to develop your model.

Chapter 10
Final Thoughts

Final Thoughts

Many people claim that the Universal Verification Methodology is hard to learn. I happen to disagree with that statement. Keep in mind that UVM is not a new language to learn; instead, it is a library of classes written in SystemVerilog to enable a common methodology. With that thought in mind, learning UVM should become less daunting.

The goal of this book was to introduce you to UVM. Hopefully, the framework is now in place for you begin to develop your own UVM environments. As you run into issues, take advantage of the resources available to you. The User's Guide and the Reference Guide are available to you in your installation. Read the source code and step through it with your simulator as you debug your code. The more you learn the library, the more efficient you will become in adapting it to your needs. There is a UVM Forum on http://www.accellera.org/community/uvm of which you can take advantage, and there are resources available through your EDA vendor if you need additional assistance.

Finally, please feel free to contact us for help! Verilab specializes in assisting customers with all facets of verification needs. Visit our website at http://www.verilab.com for more information.

Chapter 11
Appendix

RTL

```
module pipe( clk,
             rst_n,
             i_cf,
             i_en,
             i_data0,
             i_data1,
             o_data0,
             o_data1
           );

   input          clk;
   input          rst_n;
   input   [1:0]  i_cf;
   input          i_en;
   input  [15:0]  i_data0;
   input  [15:0]  i_data1;

   output [15:0]  o_data0;
   output [15:0]  o_data1;

   wire        clk;
   wire        rst_n;
   wire  [1:0] i_cf;
   wire        i_en;
   wire [15:0] i_data0;
   wire [15:0] i_data1;

   reg  [15:0] o_data0;
   reg  [15:0] o_data1;

   reg  [15:0] data_0;
   reg  [15:0] data_1;

   //Store the input data and check to see if it is
   //16'h0000 or 16'hFFFF
   //If not, multiply by correction factor

   always @(posedge clk) begin
      if(!rst_n) begin
         data_0 <= 16'h0000;
         data_1 <= 16'h0000;
      end
      else begin
         if(i_en) begin
            if((i_data0 == 16'h0000) || (i_data0 == 16'hFFFF)) begin
               data_0 <= i_data0;
            end
            else begin
               data_0 <= i_data0 * i_cf;
            end
            if((i_data1 == 16'h0000) || (i_data1 == 16'hFFFF)) begin
               data_1 <= i_data1;
            end
            else begin
               data_1 <= i_data1 * i_cf;
            end
```

```
      end
    end
  end

  always @(posedge clk) begin
    o_data0 <= data_0;
    o_data1 <= data_1;
  end
endmodule
```

DATA PACKET

```
class data_packet extends uvm_sequence_item;
  rand bit  [1:0] cf;
  rand bit        enable;
  rand bit [15:0] data_in0;
  rand bit [15:0] data_in1;
  rand bit [15:0] data_out0;
  rand bit [15:0] data_out1;
  rand int        delay;

  constraint timing {delay inside {[0:5]};}

  `uvm_object_utils_begin(data_packet)
    `uvm_field_int(cf,        UVM_DEFAULT)
    `uvm_field_int(enable,    UVM_DEFAULT)
    `uvm_field_int(data_in0,  UVM_DEFAULT)
    `uvm_field_int(data_in1,  UVM_DEFAULT)
    `uvm_field_int(data_out0, UVM_DEFAULT)
    `uvm_field_int(data_out1, UVM_DEFAULT)
    `uvm_field_int(delay,     UVM_DEFAULT)
  `uvm_object_utils_end

  function new(string name = "data_packet");
    super.new(name);
  endfunction: new

  virtual task displayAll( );
    `uvm_info("DP", $sformatf("cf = %0h enable = %0b data_in0 = %0h data_in1 = %0h
                    data_out0 = %0h data_out1 = %0h delay = %0d",
                    cf, enable, data_in0, data_in1, data_out0, data_out1,
                    delay), UVM_LOW)
  endtask: displayAll

endclass: data_packet
```

PIPE DRIVER

```
class pipe_driver extends uvm_driver #(data_packet);
   virtual pipe_if vif;

   `uvm_component_utils(pipe_driver)

   function new(string name, uvm_component parent);
      super.new(name, parent);
   endfunction: new

   function void build_phase(uvm_phase phase);
      super.build_phase(phase);
      if(!uvm_config_db#(virtual pipe_if)::get(this, "", "in_intf",
                                vif))
         `uvm_fatal("NOVIF", {"virtual interface must be set for: ",
                    get_full_name( ), ".vif"})
      `uvm_info(get_full_name( ), "Build stage complete.", UVM_LOW)
   endfunction

   virtual task run_phase(uvm_phase phase);
      fork
         reset( );
         get_and_drive( );
      join
   endtask: run_phase

   virtual task reset( );
      forever begin
         @(negedge vif.rst_n);
         `uvm_info(get_type_name( ), "Resetting signals ", UVM_LOW)
         vif.cf = 2'b0;
         vif.data_in0 = 16'b0;
         vif.data_in1 = 16'b0;
         vif.enable = 1'b0;
      end
   endtask: reset

   virtual task get_and_drive( );
      forever begin
         @(posedge vif.rst_n);
         while(vif.rst_n != 1'b0) begin
            seq_item_port.get_next_item(req);
            drive_packet(req);
            seq_item_port.item_done( );
         end
      end
   endtask: get_and_drive

   virtual task drive_packet(data_packet pkt);
      vif.enable = 1'b0;
      repeat(pkt.delay) @(posedge vif.clk);
      vif.enable = pkt.enable;
      vif.cf = pkt.cf;
      vif.data_in0 = pkt.data_in0;
      vif.data_in1 = pkt.data_in1;
```

```
      @(posedge vif.clk);
      vif.enable = 1'b0;
   endtask

endclass:pipe_driver
```

AHB DRIVER

```
ahb_driver extends uvm_driver #(ahb_transfer);

   virtual ahb_if vif;
   string  ahb_intf;

   typedef enum bit [2:0] {SINGLE, INCR, WRAP4, INCR4, WRAP8, INCR8,
                           WRAP16, INCR16} hburst_t;
   typedef enum bit [2:0] {HSIZE_8, HSIZE_16, HSIZE_32} hsize_t;
   typedef enum bit [1:0] {IDLE, BUSY, NONSEQ, SEQ} htrans_t;
   typedef enum bit       {OKAY, ERROR} hresp_t;
   typedef enum bit       {READ, WRITE} hwrite_t;
   typedef enum bit       {NOT_READY, READY} hready_t;

   mailbox #(ahb_transfer) address_box = new(1);
   mailbox #(ahb_transfer) data_box = new(1);

   function new(string name, uvm_component parent);
      super.new(name, parent);
   endfunction: new


   virtual function void build_phase(uvm_phase phase);
      super.build_phase(phase);

      if(!uvm_config_db#(virtual ahb_if)::get(this, "", "ahb_intf",
                                               vif))
         `uvm_fatal("NOVIF", {"virtual interface must be set for: ",
         get_full_name( ), ".vif"});

      `uvm_info(get_full_name( ), "Build phase complete.", UVM_LOW)
    endfunction: build_phase

    virtual task run_phase(uvm_phase phase);
       fork
          reset( );
          get_and_drive( );
          address_phase( );
          data_phase( );
       join
    endtask: run_phase

    virtual task reset( );
       @(negedge vif.hresetn);
       `uvm_info(get_type_name( ), "RESETTING SYSTEM ", UVM_LOW)
       vif.hsel <= 1'b0;
       vif.haddr <= 32'b0;
       vif.hwrite <= READ;
       vif.hsize <= HSIZE_8;
       vif.hburst <= SINGLE;
```

```
    vif.hprot <= 4'b0;
    vif.htrans <= IDLE;
    vif.hmastlock <= 1'b0;
    vif.hready <= READY;
    vif.hwdata <= 32'b0;
    `uvm_info(get_type_name( ), "DONE RESETTING SYSTEM ", UVM_LOW)
endtask: reset

virtual task get_and_drive( );
    ahb_transfer pkt;
    forever begin
        @(posedge vif.hresetn);
        `uvm_info(get_type_name( ), " DRIVING SYSTEM ", UVM_LOW)
        while(vif.hresetn != 1'b0) begin
            seq_item_port.get_next_item(req);
            $cast(pkt, req.clone( ));
            address_box.put(pkt);
            seq_item_port.item_done( );
        end
    end
endtask: get_and_drive

virtual task address_phase( );
    ahb_transfer addr_pkt;
    forever begin
        vif.htrans <= IDLE;
        address_box.get(addr_pkt);
        while(vif.hready == NOT_READY) begin
            @(posedge vif.hclk);
        end

        vif.haddr <= addr_pkt.haddr;
        vif.hburst <= addr_pkt.hburst;
        vif.hmastlock <= addr_pkt.hmastlock;
        vif.hprot <= addr_pkt.hprot;
        vif.hsize <= addr_pkt.hsize;
        vif.hwrite <= addr_pkt.hwrite;
        vif.htrans <= addr_pkt.htrans;
        vif.hsel <= 1'b1;  /* Only one slave */


        @(posedge vif.hclk);
            data_box.put(addr_pkt);
    end
endtask: address_phase

virtual task data_phase( );
    ahb_transfer data_pkt;
    forever begin
        data_box.get(data_pkt);

        if(data_pkt.hwrite == WRITE) begin
            vif.hwdata <= data_pkt.hwdata;
        `uvm_info(get_type_name( ), $sformatf("WRITING DATA: %0h at
                  %0t",data_pkt.hwdata, $time), UVM_LOW)

            while(vif.hready  == NOT_READY) begin
                @(posedge vif.hclk);
            end
        end
```

```
         if(data_pkt.hwrite == READ) begin
            @(posedge vif.hclk);
            while(vif.hready == NOT_READY) begin
               @(posedge vif.hclk);
            end
            data_pkt.hrdata = vif.hrdata;
           `uvm_info(get_type_name( ), $sformatf("READING DATA: %0h
                    at %0t",vif.hrdata, $time), UVM_LOW)
         end
         seq_item_port.put(data_pkt);
      end
   endtask: data_phase

endclass: ahb_driver
```

SEQUENCERS

```
class ahb_sequencer extends uvm_sequencer #(ahb_transfer);

   `uvm_sequencer_utils(ahb_sequencer)

   function new(string name, uvm_component parent);
      super.new(name, parent);
   endfunction: new

endclass: ahb_sequencer

class pipe_sequencer extends uvm_sequencer #(data_packet);

   `uvm_sequencer_utils(pipe_sequencer)

   function new(string name, uvm_component parent);
      super.new(name, parent);
   endfunction: new

endclass: pipe_sequencer
```

SEQUENCE LIBRARY

```
class random_sequence extends uvm_sequence #(data_packet);
   `uvm_object_utils(random_sequence)

   function new(string name = "random_sequence");
      super.new(name);
   endfunction:new

   virtual task body( );
      `uvm_do(req);
   endtask: body

endclass: random_sequence

class data0_sequence extends uvm_sequence #(data_packet);
```

```
    `uvm_object_utils(data0_sequence)

    function new(string name = "data0_sequence");
       super.new(name);
    endfunction:new

    virtual task body( );
       `uvm_do_with(req, {req.data_in0 == 16'h0;})
    endtask: body

endclass: data0_sequence


class data1_sequence extends uvm_sequence #(data_packet);
    `uvm_object_utils(data1_sequence)
    function new(string name = "data1_sequence");
       super.new(name);
    endfunction:new

    virtual task body( );
       req = data_packet::type_id::create("req");
       start_item(req);
       assert(req.randomize( ) with {data_in1 == 'hffff;});
       finish_item(req);
    endtask: body

endclass: data1_sequence

class many_random_sequence extends uvm_sequence #(data_packet);
    rand int loop;

    constraint limit {loop inside {[5:10]};}
    `uvm_object_utils(many_random_sequence)

    function new(string name = "many_random_sequence");
       super.new(name);
    endfunction:new

    virtual task body( );
       for(int i = 0; i < loop; i++) begin
         `uvm_do(req);
       end
    endtask: body

endclass: many_random_sequence
```

PIPE MONITOR

```
class pipe_monitor extends uvm_monitor;
    virtual pipe_if vif;
    string monitor_intf;
    int num_pkts;

    uvm_analysis_port #(data_packet) item_collected_port;
    data_packet data_collected;
    data_packet data_clone;
```

```
   `uvm_component_utils(pipe_monitor)

   function new(string name, uvm_component parent);
      super.new(name, parent);
   endfunction: new

   function void build_phase(uvm_phase phase);
      super.build_phase(phase);
      if(!uvm_config_db#(string)::get(this, "", "monitor_intf",
                                      monitor_intf))
         `uvm_fatal("NOSTRING", {"Need interface name for: ",
         get_full_name( ), ".monitor_intf"})

      `uvm_info(get_type_name( ), $sformatf("INTERFACE USED = %0s",
                                             monitor_intf), UVM_LOW)

      if(!uvm_config_db#(virtual pipe_if)::get(this, "",
                                               monitor_intf, vif))
         `uvm_fatal("NOVIF", {"virtual interface must be set for: ",
         get_full_name( ), ".vif"})

      item_collected_port = new("item_collected_port", this);
      data_collected =
      data_packet::type_id::create("data_collected");

      data_clone = data_packet::type_id::create("data_clone");

      `uvm_info(get_full_name( ), "Build stage complete.", UVM_LOW)
   endfunction: build_phase

   virtual task run_phase(uvm_phase phase);
      collect_data( );
   endtask: run_phase

   virtual task collect_data( );
      forever begin
         wait(vif.enable)
         data_collected.cf = vif.cf;
         data_collected.data_in0 = vif.data_in0;
         data_collected.data_in1 = vif.data_in1;
         repeat(3) @(posedge vif.clk);
         data_collected.data_out0 = vif.data_out0;
         data_collected.data_out1 = vif.data_out1;
         $cast(data_clone, data_collected.clone( ));
         item_collected_port.write(data_clone);
         num_pkts++;
      end
   endtask: collect_data

   virtual function void report_phase(uvm_phase phase);
      `uvm_info(get_type_name( ), $sformatf("REPORT: COLLECTED
      PACKETS = %0d", num_pkts), UVM_LOW)
   endfunction: report_phase


endclass: pipe_monitor
```

PIPE AGENT AND ENV

```
class pipe_agent extends uvm_agent;
   protected uvm_active_passive_enum is_active = UVM_ACTIVE;

   pipe_sequencer sequencer;
   pipe_driver    driver;
   pipe_monitor   monitor;

   `uvm_component_utils_begin(pipe_agent)
      `uvm_field_enum(uvm_active_passive_enum, is_active, UVM_ALL_ON)
   `uvm_component_utils_end

   function new(string name, uvm_component parent);
      super.new(name, parent);
   endfunction

   function void build_phase(uvm_phase phase);
      super.build_phase(phase);
      if(is_active == UVM_ACTIVE) begin
         sequencer = pipe_sequencer::type_id::create("sequencer",
         this);

         driver = pipe_driver::type_id::create("driver", this);
      end

      monitor = pipe_monitor::type_id::create("monitor", this);

      `uvm_info(get_full_name( ), "Build stage complete.", UVM_LOW)
   endfunction: build_phase

   function void connect_phase(uvm_phase phase);
      if(is_active == UVM_ACTIVE)
         driver.seq_item_port.connect(sequencer.seq_item_export);
      `uvm_info(get_full_name( ), "Connect stage complete.", UVM_LOW)
   endfunction: connect_phase
endclass: pipe_agent

class pipe_env extends uvm_env;

   pipe_agent agent;

   `uvm_component_utils(pipe_env)

   function new(string name, uvm_component parent);
      super.new(name, parent);
   endfunction

   function void build_phase(uvm_phase phase);
      super.build_phase(phase);

      agent = pipe_agent::type_id::create("agent", this);

      `uvm_info(get_full_name( ), "Build stage complete.", UVM_LOW)
   endfunction: build_phase

endclass: pipe_env
```

PIPE SCOREBOARD

```
class pipe_scoreboard extends uvm_scoreboard;
   uvm_tlm_analysis_fifo #(data_packet) input_packets_collected;
   uvm_tlm_analysis_fifo #(data_packet) output_packets_collected;

   data_packet input_packet;
   data_packet output_packet;

   `uvm_component_utils(pipe_scoreboard)

   function new(string name, uvm_component parent);
      super.new(name, parent);
   endfunction: new

   function void build_phase(uvm_phase phase);
      super.build_phase(phase);

      input_packets_collected = new("input_packets_collected", this);
      output_packets_collected = new("output_packets_collected",
      this);

      input_packet = data_packet::type_id::create("input_packet");
      output_packet = data_packet::type_id::create("output_packet");

      `uvm_info(get_full_name( ), "Build stage complete.", UVM_LOW)
   endfunction: build_phase

   virtual task run_phase(uvm_phase phase);
      super.run_phase(phase);
       watcher( );
   endtask: run_phase

   virtual task watcher( );
      forever begin
         input_packets_collected.get(input_packet);
         output_packets_collected.get(output_packet);
         compare_data( );
      end
   endtask: watcher

   virtual task compare_data( );
      bit [15:0] exp_data0;
      bit [15:0] exp_data1;

      if((input_packet.data_in0 == 16'h0000) ||
         (input_packet.data_in0 == 16'hffff))
         exp_data0 = input_packet.data_in0;
      else
         exp_data0 = input_packet.data_in0 * input_packet.cf;

      if((input_packet.data_in1 == 16'h0000) ||
         (input_packet.data_in1 == 16'hffff))
         exp_data1 = input_packet.data_in1;
      else
         exp_data1 = input_packet.data_in1 * input_packet.cf;

      if(exp_data0 != output_packet.data_out0)
         `uvm_error(get_type_name( ), $sformatf("Actual output data
```

```
        %0h does not match expected %0h",output_packet.data_out0,
        exp_data0))

    if(exp_data1 != output_packet.data_out1)
       `uvm_error(get_type_name( ), $sformatf("Actual output data
        %0h does not match expected %0h", output_packet.data_out1,
        exp_data1))

  endtask:compare_data

endclass: pipe_scoreboard
```

PIPE COVERAGE

```
class pipe_coverage extends uvm_subscriber #(data_packet);

  data_packet pkt;
  int count;

  `uvm_component_utils(pipe_coverage)

  covergroup cg;
     option.per_instance = 1;
     cov_cf:    coverpoint pkt.cf;
     cov_en:    coverpoint pkt.enable;
     cov_ino:   coverpoint pkt.data_in0;
     cov_in1:   coverpoint pkt.data_in1;
     cov_out0:  coverpoint pkt.data_out0;
     cov_out1:  coverpoint pkt.data_out1;
     cov_del:   coverpoint pkt.delay;
  endgroup: cg

  function new(string name, uvm_component parent);
     super.new(name, parent);
     cg = new( );
  endfunction: new

  function void write(data_packet t);
     pkt = t;
     count++;
     cg.sample( );
  endfunction: write

  virtual function void extract_phase(uvm_phase phase);
     `uvm_info(get_type_name( ), $sformatf("Number of coverage
packets collected = %0d", count), UVM_LOW)
     `uvm_info(get_type_name( ), $sformatf("Current coverage  = %f",
cg.get_coverage( )), UVM_LOW)
  endfunction: extract_phase
endclass: pipe_coverage
```

TEST LIBRARY

```
class base_test extends uvm_test;
   `uvm_component_utils(base_test)

   dut_env env;
   uvm_table_printer printer;

   function new(string name, uvm_component parent);
      super.new(name, parent);
   endfunction: new

   function void build_phase(uvm_phase phase);
      super.build_phase(phase);
      env = dut_env::type_id::create("env", this);
      printer = new( );
      printer.knobs.depth = 5;
   endfunction:build_phase

   virtual function void end_of_elaboration_phase(uvm_phase phase);
      `uvm_info(get_type_name( ), $sformatf("Printing the test
       topology :\n%s", this.sprint(printer)), UVM_LOW)
   endfunction: end_of_elaboration_phase

   virtual task run_phase(uvm_phase phase);
      phase.phase_done.set_drain_time(this, 1500);
   endtask: run_phase

endclass: base_test

class random_test extends base_test;
   `uvm_component_utils(random_test)

   function new(string name, uvm_component parent);
      super.new(name, parent);
   endfunction: new

   function void build_phase(uvm_phase phase);
      super.build_phase(phase);
   endfunction: build_phase

   virtual task run_phase(uvm_phase phase);
      random_sequence seq;

      super.run_phase(phase);
      phase.raise_objection(this);
      seq = random_sequence::type_id::create("seq");
      seq.start(env.penv_in.agent.sequencer);
      phase.drop_objection(this);
   endtask: run_phase
endclass: random_test

class data0_test extends base_test;
   `uvm_component_utils(data0_test)

   function new(string name, uvm_component parent);
      super.new(name, parent);
   endfunction: new
```

```
   function void build_phase(uvm_phase phase);
      super.build_phase(phase);
   endfunction: build_phase

   virtual task run_phase(uvm_phase phase);
      data0_sequence seq;

      super.run_phase(phase);
      phase.raise_objection(this);
      seq = data0_sequence::type_id::create("seq");
      seq.start(env.penv_in.agent.sequencer);
      phase.drop_objection(this);
   endtask: run_phase
endclass: data0_test

class data1_test extends base_test;
   `uvm_component_utils(data1_test)

   function new(string name, uvm_component parent);
      super.new(name, parent);
   endfunction: new

   function void build_phase(uvm_phase phase);
      super.build_phase(phase);
   endfunction: build_phase

   virtual task run_phase(uvm_phase phase);
      data1_sequence seq;

      super.run_phase(phase);
      phase.raise_objection(this);
      seq = data1_sequence::type_id::create("seq");
      seq.start(env.penv_in.agent.sequencer);
      phase.drop_objection(this);
   endtask: run_phase
endclass: data1_test

class many_random_test extends base_test;
   `uvm_component_utils(many_random_test)

   function new(string name, uvm_component parent);
      super.new(name, parent);
   endfunction: new

   function void build_phase(uvm_phase phase);
      super.build_phase(phase);
   endfunction: build_phase

   virtual task run_phase(uvm_phase phase);
      many_random_sequence seq;

      super.run_phase(phase);
      phase.raise_objection(this);
      seq = many_random_sequence::type_id::create("seq");
      assert(seq.randomize( ));
      seq.start(env.penv_in.agent.sequencer);
      phase.drop_objection(this);
   endtask: run_phase
endclass: many_random_test
```

TOP LEVEL

```
module top;
   import uvm_pkg::*;
   import pipe_pkg::*;

   bit clk;
   bit rst_n;

   pipe_if ivif(.clk(clk), .rst_n(rst_n));
   pipe_if ovif(.clk(clk), .rst_n(rst_n));


   pipe pipe_top(.clk(clk),
                 .rst_n(rst_n),
                 .i_cf(ivif.cf),
                 .i_en(ivif.enable),
                 .i_data0(ivif.data_in0),
                 .i_data1(ivif.data_in1),
                 .o_data0(ovif.data_out0),
                 .o_data1(ovif.data_out1)
                );

   always #5 clk = ~clk;

   initial begin
       #5 rst_n = 1'b0;
      #25 rst_n = 1'b1;
   end

   assign ovif.enable = ivif.enable;

   initial begin
      uvm_config_db#(virtual pipe_if)::set(uvm_root::get( ) ,
                    "*.agent.*" , "in_intf", ivif);
      uvm_config_db#(virtual pipe_if)::set(uvm_root::get( ) ,
                     "*.monitor" , "out_intf", ovif);

      run_test( );
   end

endmodule
```

PIPE PACKAGE

```
package pipe_pkg;
   import uvm_pkg::*;
   `include "uvm_macros.svh"
   `include "data_packet.sv"
   `include "pipe_driver.sv"
   `include "pipe_monitor.sv"
   `include "pipe_sequencer.sv"
   `include "pipe_agent.sv"
   `include "pipe_scoreboard.sv"
   `include "pipe_coverage.sv"
   `include "pipe_env.sv"
   `include "dut_env.sv"
   `include "pipe_sequence_lib.sv"
   `include "test_lib.sv"
endpackage: pipe_pkg
```

REGISTER MODEL

```
class regs_control_reg extends uvm_reg;
   rand uvm_reg_field enable;
   rand uvm_reg_field mode_select;

   `uvm_object_utils(regs_control_reg)

   function new(string name = "regs_control_reg");
      super.new(name, 32, UVM_NO_COVERAGE);
   endfunction: new

   virtual function void build( );
      this.enable = uvm_reg_field::type_id::create("enable");
      this.enable.configure(this, 16, 0, "RW", 0, 1'h0, 1, 0, 0);

      this.mode_select =
      uvm_reg_field::type_id::create("mode_select");
      this.mode_select.configure(this, 16, 16, "RW", 0,
      1'h0, 1, 0, 0);
   endfunction: build
endclass: regs_control_reg

class regs_status_reg extends uvm_reg;
   rand uvm_reg_field status;

   `uvm_object_utils(regs_status_reg)

   function new(string name = "regs_status_reg");
      super.new(name, 32, UVM_NO_COVERAGE);
   endfunction: new

   virtual function void build( );
      this.status = uvm_reg_field::type_id::create("status");
      this.status.configure(this, 32, 0, "RO", 1, 1'h0, 1, 0, 0);
```

```
   endfunction: build
endclass: regs_status_reg

class dut_regs extends uvm_reg_block;
   rand regs_control_reg control_reg;
   rand regs_status_reg  status_reg;

   `uvm_object_utils(dut_regs)

   function new(string name = "dut_regs");
      super.new(name, UVM_NO_COVERAGE);
   endfunction: new

   virtual function void build( );
      this.default_map = create_map("dut_reg_map", 0, 4,
      UVM_LITTLE_ENDIAN, 0);

      this.control_reg =
      regs_control_reg::type_id::create("control_reg");

      this.control_reg.build( );
      this.control_reg.configure(this, null, "");
      this.default_map.add_reg(this.control_reg,
      `UVM_REG_ADDR_WIDTH'h04, "RW", 0);

      this.status_reg =
      regs_status_reg::type_id::create("status_reg");
      this.status_reg.build( );
      this.status_reg.configure(this, null, "");
      this.default_map.add_reg(this.status_reg,
      `UVM_REG_ADDR_WIDTH'h08, "RO", 0);

      lock_model( );
   endfunction: build


endclass: dut_regs

class reg2ahb_adapter extends uvm_reg_adapter;

   `uvm_object_utils(reg2ahb_adapter)

   function new(string name = "reg2ahb_adapter");
      super.new(name);
      provides_responses = 1;
   endfunction: new

   virtual function uvm_sequence_item reg2bus(const ref
   uvm_reg_bus_op rw);

      ahb_transfer ahb_pkt =
      ahb_transfer::type_id::create("ahb_pkt");

      assert(ahb_pkt.randomize( ));

      ahb_pkt.hwrite = (rw.kind == UVM_READ) ? READ : WRITE;
      ahb_pkt.haddr = rw.addr;

      if(rw.kind == UVM_WRITE) begin
         ahb_pkt.hwdata = rw.data;
      end
```

```
      else begin
         ahb_pkt.hrdata = rw.data;
      end
      return ahb_pkt;
   endfunction: reg2bus

   virtual function void bus2reg(uvm_sequence_item bus_item,
   ref uvm_reg_bus_op rw);

      ahb_transfer ahb_pkt;
      if(!$cast(ahb_pkt, bus_item)) begin
         `uvm_fatal("NOT AHB TYPE", "Provided bus_item is not of the
         correct type")
         return;
      end
      rw.kind = ahb_pkt.hwrite ? UVM_WRITE : UVM_READ;
      rw.addr = ahb_pkt.haddr;
      if(ahb_pkt.hwrite == UVM_WRITE) begin
         rw.data = ahb_pkt.hwdata;
      end
      else begin
         rw.data = ahb_pkt.hrdata;
      end
      rw.status = UVM_IS_OK;
   endfunction: bus2reg

endclass: reg2ahb_adapter

class dut_env extends uvm_env;

   ...
   ahb_env ahb_bus_env;
   dut_regs regmodel;
   dut_regs sb_regmodel;
   reg2ahb_adapter stim_adapter;
   reg2ahb_adapter sb_adapter;
   uvm_reg_predictor#(ahb_transfer) mpredictor;

   `uvm_component_utils(dut_env)

   function new(string name, uvm_component parent);
      super.new(name, parent);
   endfunction

   virtual function void build_phase(uvm_phase phase);
      super.build_phase(phase);

      ...
      ahb_bus_env = ahb_env::type_id::create("ahb_bus_env", this);

      regmodel = dut_regs::type_id::create("regmodel", this);
      sb_regmodel = dut_regs::type_id::create("sb_regmodel", this);

      mpredictor =
      uvm_reg_predictor#(ahb_transfer)::type_id::create("mpredictor",
      this);

      ...
   endfunction: build_phase

   virtual function void connect_phase(uvm_phase phase);
```

```
    ...

    //Connect the Register Models
    if(regmodel.get_parent( ) == null) begin
       stim_adapter =
       reg2ahb_adapter::type_id::create("stim_adapter",,
       get_full_name( ));

       regmodel.default_map.set_sequencer(
       ahb_bus_env.agent.sequencer, stim_adapter);

       regmodel.default_map.set_auto_predict(1);
    end

    if(sb_regmodel.get_parent( ) == null) begin
       sb_adapter = reg2ahb_adapter::type_id::create("sb_adapter",,
       get_full_name( ));

       sb_regmodel.default_map.set_sequencer(
       ahb_bus_env.agent.sequencer, sb_adapter);

       mpredictor.map = sb_regmodel.default_map;
       mpredictor.adapter = sb_adapter;
       sb_regmodel.default_map.set_auto_predict(0);
       sb_regmodel.default_map.set_check_on_read(1);

       ahb_bus_env.agent.monitor.item_collected_port.connect(
       mpredictor.bus_in);
    end
    ...

  endfunction: connect_phase

endclass: dut_env
```

References

Meade, K.A., Rosenberg, S. (2010). A Practical Guide to Adopting the Universal Verification Methodology (UVM). San Jose: Cadence Design Systems, Inc.

Accellera (Eds.). (2011). Universal Verification Methodology (UVM) 1.1 User's Guide.

Accellera (Eds.). (2011). UVM 1.1 Class Reference.

http://www.accellera.org

http://www.accellera.org/community/uvm

http://www.verificationacademy.com

www.ingramcontent.com/pod-product-compliance
Lightning Source LLC
Chambersburg PA
CBHW041710200326
41518CB00001B/144

9780615819976